Genetic Destinies

Genetic Destinies

Peter Little

OXFORD
UNIVERSITY PRESS

OXFORD

UNIVERSITY PRESS

Great Clarendon Street, Oxford OX2 6DP

Oxford University Press is a department of the University of Oxford.
It furthers the University's objective of excellence in research, scholarship, and education by
publishing worldwide in

Oxford New York

Auckland Bangkok Buenos Aires Cape Town Chennai Dar es Salaam Delhi Hong Kong Istanbul
Karachi Kolkata Kuala Lumpur Madrid Melbourne Mexico City Mumbai Nairobi São Paulo
Shanghai Singapore Taipei Tokyo Toronto

with an associated company in Berlin

Oxford is a registered trade mark of Oxford University Press in the UK and in certain other countries

Published in the United States by Oxford University Press Inc., New York

British Library Cataloguing in Publication Data
Data available

Library of Congress Cataloguing in Publication Data
Data available

ISBN 0198504543 (Hbk)

Typeset in Minion by Footnote Graphics, Warminster, Wilts
Printed in Great Britain on acid-free paper
by T.J. International, Padstow, Cornwall

To my Father, who did not live to read it;
to Natasha and Fenella,
who are the future; and to Liz,
beautiful, blue-eyed and red-haired.

Contents

Contents

Acknowledgements

This book was conceived out of conversations with my father, who made me realize that genetic research is a topic of defining interest to all well-educated individuals but that modern genetic research is too immersed in detail to be readily comprehensible to any but the specialist. Sadly, he did not live to see the results of our many conversations.

Genetic Destinies was written over a period of 3 years during a time of great change in my life and also of great sadness—the deaths of my father and my sister. I owe an unrepayable debt of gratitude to my partner, Liz Jazwinska, for her love and unquestioning support over this time. Without her, I doubt I could have faced the difficulties that we have faced together, and I certainly never could have written this book. If a scientific writer needs a muse, Liz has been my muse and she has been my keenest critic and sounding board: in every respect, I owe her an enormous debt of gratitude and love.

Many people have contributed to the ideas and themes in *Genetic Destinies*: endless conversations and arguments and to acknowledge them all would be a Sisyphean task. In particular, though, Keith Willison and Robin Brown have bravely endured hours of talk and the members of my own laboratory and my students have contributed far more than they realized in making me understand what is important and what is detail.

I would also like to thank several anonymous reviewers for some very helpful comments and Professor Paul Mathews for setting my thoughts straight on magnetic resonance imaging (MRI). The idea of using what I call 'future histories' as a device to explore the limits of genetics came to me after listening to an excellent talk given by Marty Rosenberg, of SmithKline Beecham, at a conference in Brisbane in 1999 and I am delighted to acknowledge his contribution.

Michael Rodgers, my editor at Oxford University Press, has been extraordinarily understanding in the drawn-out birth of *Genetic Destinies*, and his advice and support have been critical to me. Abbie Headon, also of the Press, has eased the many passages that the manuscript has made from England to Australia, and her help and advice are most happily acknowledged. Finally, Sarah Bunney has been a wonderful help in copyediting this book, and her suggestions and eye for detail have combined to make the writing far better than I ever could have hoped to have achieved unaided. Some tasks are too great; the remaining vagaries and errors of my scientific writing remain my responsibility, and mine alone!

Sydney
November 2001

Genes, dreams, and nightmares

In the early hours of the morning of the last day of August in the year 2020 two girls were born; this is the story of their lives, told as their future history might be. The chapters that follow contain the science of why their lives might, or might not, unfold in these ways and they are concerned with genes and how they influence our existence. As we will see, our personal futures are in our own hands but they are also in the hands of our genes: these are our Genetic Destinies.

■ The gene dream

This is the future history of the life and death of Jeanne Dream, who was born in the small town of Prosperous. Her parents, Perfect and Manley Dream, had been married for 5 years before moving to Prosperous to avoid the high taxes and crime of the City and to find a safe place to raise a family. One year before Jeanne's birth, they both attended Dr Goode's surgery, where DNA tests were run to check for unfortunate gene differences that might give rise to problems for their baby. They were also counselled as to the possible range of intelligence, hair colours, skin complexion, and major personality attributes their baby might possess.

Heartened by this information, Perfect and Manley go home and, as in the best stories, they share a wonderful meal together. The next morning they return to Dr Goode for an overnight stay in her Love Ward. Several more visits follow, at the end of which Dr Goode has taken 10 eggs from Perfect and fertilized them with sperm from Manley in an *in vitro* fertilization (IVF) operation. The fertilized eggs develop normally and at 10 days cells are removed for standard DNA testing and for possible banking of embryo stem cells. The results of these tests are presented to Perfect and Manley.

Embryo 10 looks particularly promising: it is female with a very low chance of suffering from any of the disorders caused by single genes. There is a 20 per cent chance of common cancers appearing; Alzheimer's has a 50 per cent chance, major depressive illness 10 per cent, and schizophrenia 1 per cent. The future child's personality profile is extraversion 60 per cent, neuroticism 10 per cent, agreeableness 70 per cent, conscientiousness 70 per cent, and culture 60 per cent. The only bad news is that the possibility of dyslexia, addictive personality trait, and attention-deficit disorder is 90 per cent, and there is a potential for criminal actions. But the gene counsellors advise this is 99.9 per cent unlikely to occur within the socially advantageous environment that embryo 10 will occupy. She will be red-haired, pale-skinned, and freckled—and be at risk from skin cancer.

Delighted by most of this, Perfect and Manley choose embryo 10 for stem-cell culture, and this is implanted into Perfect. The pregnancy proceeds normally: 9 months later, on 31 August 2020, little Jeanne emerges into the world after an uneventful pregnancy. Jeanne is immediately retyped for the same set of disorders to confirm the absence of DNA changes during development; all the results replicate the earlier ones. She is also DNA-tested to detect her immune-system MHC (major histocompatibility complex) type. Based on these results she is immunized against TB, measles, diphtheria, and five common cancers using vaccines profiled to her MHC gene differences. She is shown to be HIV-resistant and so is not immunized against that disease. Finally, Jeanne's DNA is tested for differences to the 155 genes that encode the proteins on the 90 per cent drug-response indicators list compiled by GlaxoSmithMerkNovartis—the GSMN-90—to ensure any future drug treatment is matched to her particular profile of variation. This information is recorded on a remote readable chip that is implanted under the skin at the back of her neck.

When Jeanne is aged two, the barbecue she is standing next to flares up and she receives burns to 45 per cent of her right arm. She is rushed to

Prosperous Hospital and treated successfully with a skin-cell culture from her bank of embryo stem cells.

Four years old, and based on her new school's recommendation, she starts a 1-year course of gene therapy against attention-deficit disorder. This is followed by dyslexia therapy using the GSMN cognitive enhancer (Iqzac) and intensive teaching therapy. The outcome is positive, but the school reports that Jeanne is very shy and so she is treated with GSMN's latest social enhancer (Zocialac). The school is delighted at the results and the rest of her schooling is uneventful: she graduates with excellent grades and enrols at Prosperous Private University.

Just before she is due to go to university, on her eighteenth birthday, Jeanne is involved in a car crash and her left arm is badly crushed. She is left in a coma for 5 days. Rushed to the Prosperous GSMN trauma unit, her GSMN-90 chip is read and she is treated with an individually profiled selection of anti-inflammatory drugs to reduce the trauma to her head. Her crushed arm is treated with GSMN's Growzac, a bone-growth enhancer, which was discovered in 2010. Because of her relationship to GSMN, she is enrolled in a trial involving one of GSMN's latest products—'developmental restarting initiators'. She is treated with a cocktail of 10 naturally occurring developmental regulators that can reprogram the genes that are being expressed in the cells of a tissue so that they develop the capacity to renew themselves.

The treatment is successful for all parts of Jeanne's arm, but her hand remains irreversibly damaged and residual nerve damage requires further treatment with two compounds developed out of GSMN's 'developmental re-orientation of potential' (DROP) programme, which is seeking to develop ways of transforming any tissue type into any other type, by reprogramming gene activity. In Jeanne's case, connective tissue cells are reprogrammed into nerve cells and the treatment of the damaged nerves is successful. Her hand, though, is in a bad way and her surgeons elect to carry out an organ transplant using a hand grown from an embryo by Embryofarm, a noted embryoid development company. The tissue match is perfect (it was selected to be so, based on matching DNA differences) and the organ grows perfectly, but Jeanne will forever have a slight mismatch of skin colour on her new hand. This is the subject of a legal case that is resolved after 2 years with Embryofarms admitting they failed to conform to their DNA difference standard for racial origin of 99 per cent purity.

Long before then, a graft from Jeanne's embryo stem-cell bank resolves the residual tissue loss on the scarred surface of her arm. In an after-note to the case, the Embryofarm surrogate mother who incubated the hand, a young woman of 18 called Jean Battler, is fired for failing to disclose an accurate gene record. The company knows this is nonsense because they DNA-typed her themselves, but it enabled them to escape some of the legal liabilities.

Jeanne starts university a year late because of her injuries and time in hospital. In her first year she shows the initial signs of alcohol abuse and is successfully treated with a combination of receptor blockers and serotonin analogues (GSMN's Banishac). She travels to Africa for her middle year to help in an anti-tsetse fly eradication programme using tsetse flies that have been behaviourally modified genetically. Wading through a swamp to rescue her best friend, she acquires a bad infection of bilharzia, but this is successfully treated with GSMN's Billyhac that had been made after the identification of the drug target (a DNA-copying protein) in the bilharzia genome project in 2002.

In 2041, Jeanne graduates with a good degree and goes to work in the pharmaceutical giant GlaxoSmithMerkNovartis as a financial analyst. She has a very successful career and social life for 15 years during which time she meets, and finally marries, a tall, wonderfully kind, blue-eyed, PhD-level GSMN gene scientist called Wilde Hope. Within 2 years, Wilde and Jeanne decide to have children and Jeanne elects for a gene-therapy enhanced pregnancy because she is 38 years old.

Jeanne undergoes IVF, just as her mother did to conceive her, but on this occasion there are only three fertilized eggs for testing. The sex of all of them is determined as male and a gene scan of the embryos with the GSMN-99 DNA chip test for 99 per cent of all gene-related conditions shows a less than 10 per cent chance of all conditions, except for four. The gene differences that predisposed to these conditions are rectified by hit-and-run gene therapy using GSMN's latest Hitandrunzac embryo gene-correction system. At Wilde's father's request, the baby's body size is augmented by 20 per cent and muscle bulk by 5 per cent, using GSMN's Bulkzac embryo gene-augmentation system. Nine months later Wilde junior is born and put down for sports education at a top private academy, Prosperous College.

Jeanne's post-natal depression is treated with Anxzac, a tenth-generation anti-depressant just released by GSMN and discovered as part of their behavioural gene programme that was initiated way back in 2005. She is DNA-tested at the same time using the new GSMN-99.9 chip. Her personal data chip, still implanted in her neck, is updated.

Now 43, Jeanne is in the prime of life; 20 years of ecstatically happy life follow. This is punctuated by the occasional visit to Dr Goode's surgery for a variety of minor infections and ailments, all treated with programmed anti-bacterial and immune-response enhancers that are individually profile-matched to her GSMN-99.9 chip and to GSMN's pathogen DNA-sequence database, which contains the gene structure of 90 per cent of all known human pathogens. Wilde takes very lucrative retirement from GSMN, and in the same year Wilde Jr takes a silver medal at the 2085 international games (he is entered into the drug-free but gene-augmented category of the strongest fastest section).

When Jeanne is aged 78, 15 years later, cognitive testing detects the first sign of Alzheimer's disease and she is treated with gene therapy to prevent the loss of brain tissue. Over the next 10 years she is treated once a year. Two cancers are identified during these routine visits and both are treated with a combination of gene therapy aimed at forcing the tumours to differentiate into non-dividing tissues and immune-system stimulators that trick the body's defences into thinking the tumour is 'non-self', triggering it to destroy the cancer. All treatment uses GSMN's standard anti-tumour drugs and is 100 per cent successful.

By the end of this period, Jeanne is 88 and her anti-Alzheimer treatment is only moderately successful. She is placed on intensive cognitive-enhancer treatment (Iqzac, once again) but also on some third-generation DROP products to restore nerve-cell division. Jeanne has noticed she can manage only two sets of tennis before her aches start to inhibit her movement; these arthritic conditions are treated at the same time using specific immune-system suppressants and bone-regrowth programmers (again DROP products): success is complete.

Another 10 years pass with continual, successful, treatment for Alzheimer's and arthritis and the usual crop of minor ailments, including two or three cancers. Jeanne's heart is becoming somewhat enlarged but this is taken care of by a programme of cardiac de-differentiators that have been available for 30 years. During these investigations, a weakening of two blood vessels is repaired by a local application of cell-growth enhancers and cells from her embryo cell bank. As time progresses, she starts to suffer from some loss of muscle cells but this is easily treated with the DROP programme drug—third-generation ReBulkzac.

Finally, on the last day of August 2140, Jeanne and Wilde lie down on their bed, together for the last time, and both swallow a single pill that contains a complex mixture of muscle relaxants, cognitive stimulators, neurotoxins, and nerve-function inhibitors. With music playing gently in the background, and hand in hand, Jeanne and Wilde gently and happily die, content in the knowledge that theirs has been a wonderful and ful-filling life, untouched by pain, diminution of abilities with age, and sure of the rightness of their decision to leave life together. Jeanne lived and died in the future, where disease and suffering were an echo of the past.

■ The gene nightmare

This is the future history of the life and death of Jean Battler, who was born in the small town of Poorsville. Her mother was called Struglin' by everyone who knew her—she could no longer remember a time when anyone had called her by her real name of Hope. Struglin' had no idea who her baby's father was—there had been several possibilities. But none of her boyfriends had stayed around long enough to find out she was pregnant, and she was not overly worried by this.

Struglin' gives birth to Jean in Poorsville district hospital and, under the public-health maintenance scheme, Jean is immediately DNA-tested for the usual set of major social-impact conditions. These show that she has a 50 per cent chance of developing Alzheimer's disease and an 80 per cent chance of breast or ovarian cancer. Struglin' is advised by her birth para-medic that Jean will never be eligible for the state medical insurance scheme. Because Struglin' has no money to cover this in any event, she really does not care. Jean's IQ is established by intelligence-gene testing as being of no greater than 80—well below the minimum 90 required for entry to the state education system.

More worryingly, the DNA tests show Jean to have a criminal predis-position for violence and the genetic para-counsellor advises that the state would seek preventative imprisonment in a foster home. Distraught, Struglin', who is deep in the depths of post-natal depression, slashes Jean's and her own wrists with a knife stolen from the kitchen. A chance saves Jean, but Struglin' dies, certain she would be forgiven the terrible crime she had committed against her baby to save her from an inevitable life of suffering and deprivation.

Over the next 15 or so years, Jean is brought up and educated, poorly, in a state institution for the genetically underprivileged (SIGU). She suffers from dyslexia and attention-deficit disorder, and remains functionally illiterate. In 2035, after a military *coup d'état* and victory by white supremacists, all occupants of SIGUs are tested for non-Caucasian DNA ancestry, using a battery of 24 DNA differences. Jean has 13 DNA differences that are non-Caucasian and, because this number is greater than 50 per cent, she is forcibly expelled to the United People's Country.

Aged just 15 and destitute, Jean joins Embryofarms as a baby-part surrogate and is paid to give birth to a brain, a headless baby, and a hand—all made using Embryofarm's standard IVF and blocking gene therapy. The embryoids are used for spare-part surgery but the company summarily fires Jean for non-disclosure of her non-Caucasian DNA test results. She survives on social payments for 2 years, but following a government decree that all citizens must undergo compulsory DNA profiling for DNA differences likely to disadvantage society, Jean is declared genetically unsuitable for reproduction by an increasingly right-wing government of United People's Country and is forcibly sterilized. Her profile includes the chance of common cancers at 20 per cent, Alzheimer's at 50 per cent, major depressive illness at 10 per cent, schizophrenia at 1 per cent, and dyslexia, addictive personality trait, attention-deficit disorder, and general criminality all 90 per cent—and grounds for sterilization.

Jean lives a very poor existence, surviving on her wits, some drug-dealing and stealing, and when times are hard she sells herself to men at the local rail station. Her addiction to drugs and alcohol increases and she is frequently imprisoned for her activities. She is regularly treated for a whole range of sexually transmitted diseases using twentieth-century antibiotics, which are predominantly ineffective because of multiple resistances in the disease organisms, and finally becomes infected with HIV. The state refuses her any further treatment because her DNA profile suggests that the incubation period before AIDS becomes life-threatening is greater than 4 years and therefore deemed to be outside of the 3-year minimum life-expectancy period policy (3MEPP) that triggers treatment under the medical rationing programme for the genetically disadvantaged started in the early twenty-first century.

Jean is now aged 26 but looks like a 56-year-old, her good looks having gone. In 2048, she is diagnosed as having a slow-growing brain tumour, which causes unpredictable blackouts. This is not treated under the 3MEPP. Sleeping in a derelict building, she catches TB and is found by the police, collapsed and half dead on the street, and is taken to the poor hospital where she is isolated from all human contact because she is harbouring a multi-drug-resistant TB strain. Why she does not die, no one can tell: but she doesn't. She finally emerges from hospital in late August just in time to be caught in the cloud of infectious particles released from a missile containing a viral warfare agent that has been fired from the adjacent United People's Country. The agent had been genetically modified to target only non-Caucasians and 5 days later, in great pain and blind and deaf from the nerve destruction caused by the deliberately designed tissue specificity of the warfare agent, Jean dies—on the last day of August 2048.

Jean lived and died in the future, where disease and suffering were the results of nature and malign human influence.

■ A broad stroke and the devil in detail

The dream and the nightmare are, of course, fictions but I have used them to describe most of what is good and what is bad about the uses of our understanding of genes and how they influence human lives. Parts of both dreams are already scientific reality and other parts are unlikely to become so, but there are no scientifically preposterous events: each dream includes some of our darkest fears as well as our most elemental of hopes.

My ambition in this book is to discuss what sorts of knowledge we have about genes and how they work and, knowing this, what sorts of information we may possess about our futures, in our futures. This is a very substantial ambition because human beings are creatures of quite incredible complexity and it would be far too easy to discuss too much and to lose the perspective of our knowledge in a mass of detail: in science, to know too much can sometimes be as bad as to know too little. I hope that after reading this book you will emerge with an understanding and context for the types of scientific knowledge that we have today and will then be in a better position to try and assess which of the future lives—Jeanne's or Jean's—is more likely to correspond to reality. Perhaps, knowing of the possible routes to the dystopia of Jean Battler we can avoid them, and knowing the routes to the utopia of Jeanne Dream we can achieve them. Even if we fail in this greater ambition, I hope my narrative will succeed at a more mundane level because genes are increasingly impacting on our lives and our society. I hope this book will inform your views of these events and perhaps alleviate some of the fears and concerns that people naturally experience when they are exposed to the incomprehensible jargon of genetics and the extraordinary power and beauty of genes.

We will travel a long way together through the lands of genetics: the sights we will see, explore, and discuss will be some of the most important highlights. But I do not pretend for a second that we will touch most of these as anything other than a tourist: we will visit briefly and pass on. This approach has many dangers. There is another statement about science, 'the devil is in the detail', which is based on the understanding that a broad view is insufficient to understand the myriad properties of genes and proteins and their interactions that ultimately define how they function. The role of the geneticist in this is to research deeply, seeking to develop a detailed understanding. The picture of genetics is filled with fine detail, but we will ignore this to concentrate on the larger view: it is important to keep this understanding firmly in the forefront of our minds as we discuss genetics together.

The path ahead

The separate lives of Jeanne and Jean are entwined with the malign and benign influences of genes and if we are to understand what is fiction and what is fantasy in these stories, we must understand what it is that genes do and how they influence our lives. The stories of Jeanne and Jean are also stories of how knowledge of genes can be used in ways that can be a great force for evil as well as for good. Genes are paradoxical: not only do they make us all very similar, they also make us different. In spite of this, however, many human differences are not due to the effects of genes at all. This is the paradox we must explore if we are to understand the stories of Jeanne and Jean.

So there is a path to follow through this book: the path will lead us to understanding the reality of Jeanne's and Jean's stories and, as the path winds its way in front of us, we will use it to explore the science of genes. First, we must explore what genes are and what role they have in making us human beings: this we will do in Chapter 2, where I discuss the most fundamental functions genes have in all living creatures and lay the basis for understanding why all human beings are so recognizably similar. The paradox of genes is that not only do they make us look similar, they also contribute to the differences between us; why this should be so is the topic for Chapter 3. This understanding begs the question as to why there should be gene differences between people; in Chapter 4 the distant history of human beings supplies the unlikely answer to this most critical question.

Chapter 5 focuses on gene differences and being different and the impact this has, and must have, on our lives and human societies. Differences between individuals and groups of individuals form a central feature of our existence, and so in Chapter 5 I look with a carefully dispassionate eye at the most socially contentious issues of 'race' and gene differences. Within all areas of human life, we are perhaps most uncomfortable with the idea that gene differences might influence our personalities and so in Chapter 6 I first discuss the role of genes in making the human brain and the human mind and then in Chapters 7 and 8 I turn to the possible influence of gene differences on our personalities and on our intelligence.

Can we ever escape from the gene's power? Around 30 years ago this was the domain of science fiction because although we knew that humans were made by the labour of genes we had little understanding of how genes worked. The past 30 years have witnessed a revolution in abilities and understanding. Recently the first glimmer of evidence that we can indeed alter genes within human beings has emerged, so I discuss in Chapter 9 what is possible, what is impossible, and what may be possible. In Chapter 10, drawing on the previous discussions, I return to the Dream and the Nightmare and retell the stories, word for word, pausing to apply our new knowledge. This retelling leads to the final chapter where I look carefully at the basis and reality of the fears that surround genetics and its applications. My ambition is that by confronting these fears with the understanding that we have developed together, they will be allayed and the true role of genes and genetics within our own private Genetic Destinies will emerge.

Words, jargon, and simplicity

I have been a professional geneticist for 25 years of my life and so am well aware that impenetrable jargon and a huge body of information surround the subject I research and teach. As is the case, I suspect, in many professions, some of the jargon that geneticists use seems deliberately designed to exclude outsiders. I am certain this is avoidable: in the discussions that follow, I have intentionally tried to use simple terms and words, and this has necessitated giving new and non-technical names to many of the key features of genes and their functions.

This is important in that it tries to remove the hindrance that technical names introduce to the comprehension of complicated events: for example, in the discussion of the control of genes, I use the term 'gene regulators' to describe proteins that control when a gene is used in a cell. Both words are intelligible and one might even guess, correctly, that 'gene regulators' somehow regulate genes; the technical name for these proteins, however, is 'transcription factors', which is pretty meaningless to those not in the know.

I feel that the need to think about names detracts from the intellectual effort to follow the sometimes complicated chain of logic and events that make up the gene's world. I hope that you do not feel patronized by my decision and at the end of the book I provide a short 'reverse glossary' of the simple terms I use, made complex.

Finally, even a short glance at my reverse glossary will start to explain an obvious omission in my writing; nowhere do I list the many sources of information that I have used to develop the ideas and discoveries that I discuss. The omission is deliberate and inevitable. Most of the science contained within Chapter 3 onwards is at the very cutting edge of genetics and is not yet contained even in university student-level textbooks; instead, the sources of information I have used are almost entirely scientific papers. These papers are written by professional scientists for professional scientists: they directly confront the devil in the detail and require a deep knowledge of genes to be comprehensible. I have no doubt that most non-specialists would be able to get something from many of these papers but I do not believe that the effort is worth the gain and the chances of profound misunderstanding are high. The biological reality of how genes work and behave is often counter-intuitive and so a non-specialist reader is poorly equipped to make use of the combination of jargon, knowledge, and technical familiarity that make up the best scientific papers.

I am sad that non-specialists cannot appreciate the skill and excitement of the very edges of scientific research and one of my main motivations in writing *Genetic Destinies* was precisely to bring this edge to a non-specialist audience. I hope you do not feel patronized by the omission of further reading: that is far from my intention.

Why are we all different human beings?

In reading the future histories of Jean and Jeanne, perhaps the first impression that the stories make is how different their lives are: different events, different fates, and different outcomes. This obscures a point that is so obvious that it hardly needs to be stated: Jean and Jeanne are both human beings and because of this their lives have far more in common than their histories might suggest. Perhaps it would help to imagine their stories if Jean had been born a whale and Jeanne a human being, and then to think how great would have been the differences in their lives.

Why make this seemingly obvious point? Because it highlights what is fundamental to all of our views about human beings: we have an immediate and unconditional ability to recognize ourselves and each other as being human but we also accept that all human beings look recognizably different. Put in another way, we all understand that human beings have a similar, obviously human body form—two arms, two legs, and one head. But at the same time humans are very different in innumerable detail—red hair or brown, large or small, light or dark skin. As we will see, the cause of this strange mixture of similarity and dissimilarity lies in two influences—in our genes and our lifestyles.

In the chapters that follow I will discuss the effects of our genes on the differences in all of our lives, but before we can do that we need to look at how genes make each one of us a human being, just as genes make the myriad of other species of animals or plants. As we will see, and indeed already unthinkingly accept, this is genetic determinism of the most rigid kind, but the paradox is that genes also contribute to the differences between us in a way that makes us unique to an almost unbelievable extent.

So what is it that we need to understand about genes to be able to appreciate this paradox? Surprisingly we do not need to understand the stunningly complicated world of genes in minute detail but instead we have to appreciate and discuss three key ideas:

- Genes store information: they contain the information to make proteins, which are the chemical building blocks of our bodies.
- Genes and the body plan: they define the architecture of the body.
- Genes and differences: all human beings have the same set of genes but they can be subtly different in each one of us, making us unique.

■ Genes store information: what we need to know

Jean and Jeanne are both human beings, so we can start by focusing on two fundamental questions: what makes Jean and Jeanne human beings and how does it come about that they were born as human babies of human parents? The simple answer is that genes are responsible for both events.

If you look at yourself in the mirror, what you see is predominantly protein or is made by the actions of proteins. We are not sure how many different proteins make up the human body but there are probably more than 40 000 and fewer than 120 000; subtle chemical modification of many proteins probably increased this number by perhaps three times. Somehow all of these proteins have to work together to make up the million million (trillion) cells in a human being. These in turn work together to make up the different tissues of our bodies, which co-operate to make us the human beings we are. As if this was not a complicated enough problem, the whole intricate system has to be able to start as a single fertilized egg and become a new-born baby in just 9 months.

This is an extraordinary feat of organization and produces a human being of incredible complexity: no human-built structure can come close to the intricacies of our own bodies. If this was not extraordinary enough, the body, unlike any creation we humans have ever made, is able to reproduce itself, just as all living things do, over and over again down countless generations with complete fidelity and accuracy.

The biological organization of this remarkable series of abilities is gradually emerging in an astonishing act of self-knowledge with its origins only within the past 200 years, central to which is an understanding of how the information necessary to make a living creature can be stored and utilized. Possession of stored information not only enables all living organisms to develop, grow, and reproduce in each present generation but also enabled ancestral life to do so during the past 3000 million years. How this quite extraordinary feat is achieved is what I will now discuss, for it is the exquisite ability of all living organisms to reproduce their own kind that is the foundation of our understanding of how the same process can also generate our own enormous diversity.

The code in the ladder

The central discovery of our own biology was the understanding that each cell in our body contains an independent store of information, which can be read so that the right proteins are made in the right cells at the right time; understanding the control of the release of this information is in turn the key to knowledge of how our bodies can be assembled. How this is achieved is intimately tied up with the chemical within which information is stored—DNA.

If we are to discuss genes, and how they influence human lives, we do not have to have a detailed view of how genes work, even though research into genes and DNA is very much a subject where, as I already have pointed out, the devil is indeed in the detail. Instead, what we have to discuss are the broad principles that can be applied to develop our understanding, and central to this is the way DNA stores the information to make proteins and the way this information is 'read' when it is needed.

Proteins are huge molecules but fortunately all are made up of just 20 different building blocks called 'amino acids'. What makes a protein have its specific properties is the exact order of the blocks along the long protein chain and there may be as many as 4000 of these strung together like beads of a necklace; every protein of a particular type will have exactly the same blocks in the same order and different proteins will have different orders. So the information the cell needs to make a given protein is simply the order and identity of the blocks; this is what is stored in the DNA molecule.

To understand how DNA can store this, you have to know a little about the structure of DNA. Imagine a very long ladder: take one end and fix it to the ceiling and then twist the other end so the ladder turns into a spiral attached to the ceiling; that is the structure of DNA. It contains two strands coiled around each other in a spiral or helix—hence the double helix. Now forget about the spiral, just think of the ladder, untwisted. The rails of the ladder are rather uninteresting parts of the molecule that simply serve as a support for the rungs, which, as we shall see, are the most important part of DNA. Each rung is made of just two out of four possible chemicals—A, G, C, and T. If we think of our ladder again and look just at a single rung, then one chemical is attached to the rail on the left side and a second on the right, so that each rung is made of a pair of chemicals. But there is a simple rule that always applies: if an A is on one side, a T must always be attached to the opposite rail, and a G must always be opposite a C; no other combinations are allowed. The chemicals are attached to the rails, but not to each other, so if you were to pull the DNA rails apart, then the ladder would break neatly into two halves, with one of each pair of chemicals, half rungs, still firmly attached to the rail. These rules mean that DNA is a very simple molecule, but as there are more than 3000 million (3 billion) rungs, it is also huge!

The actual order of the rungs down the ladder is very important and it is the key to understanding how DNA can store the order of amino acid 'building blocks' of a protein. The order of rungs is actually a code that is 'read' by looking at the composition of three adjacent rungs; the identity of each of the 20 amino acids is coded by a different three-rung code and in several cases the same amino acid is actually coded by several different three-rung code words. The chain of amino acids in a protein is simply encoded by a continuous sequence of code words, so a protein with 100 amino acid building blocks can be encoded by 300 rungs of DNA—the 'gene' that encodes the protein. There are probably fewer than 50 000 genes in human DNA but we cannot be sure of the exact number until we have finished establishing the exact rung order of all of our DNA—a huge undertaking known as the Human Genome Project.

The Human Genome Project

The sight of President Clinton and Prime Minister Blair announcing the decoding of human DNA will remain a formative moment for many of us; in reality, what this announcement signified was just a major waypoint on the road to establishing the rung order of all 3000 million rungs of human DNA—collectively called the Human Genome in the jargon of genetics. Why so much excitement? Because now scientists can set about reading the coded information in our own DNA and, by doing this, establish the number and identity of every protein a human cell can make. In short, we will, within 2 or 3 years, have the parts list of a human being.

A parts list, though, is not the same as either a blueprint or even an instruction manual as to how all the parts go together to make a human, and this point is best kept very clearly in mind in relation to some of the more extravagant claims for this epochal project. Why is the parts list insufficient? Simply because not all parts are used in all cells all of the time; indeed, the timing and location with which DNA's encoded information is read is absolutely central to the control of all of the multitude of events that are required to make a human being and to enable he or she to survive as an adult.

How the cells in our bodies achieve the carefully controlled release of information stored in our DNA is an unfolding story of huge complexity, one upon which geneticists are presently working with extraordinary tenacity. The key to this research is a group of proteins whose function is to control the flow of information out of a gene—the gene regulators.

Genes and regulators

The coded information stored in the DNA of a gene is not read directly by the cell; instead, proteins make a copy of the gene into 'messenger RNA' that looks like the DNA ladder split down the middle of the rungs. Even though it has only one rail with half rungs attached, the coded information is preserved and it is called the 'messenger' RNA because it carries the code to the appropriate region of the cell where it is read and, quite literally, 'translated' into new proteins. Controlling which genes are copied into messenger RNA is the key to how cells make the appropriate proteins at the appropriate times, because although every cell has the same genes only the necessary genes make proteins in each cell and the remaining continue silent.

How the cell controls which genes are used and which are not is gradually being understood. Near every gene, frequently at one end, there are special regions of DNA whose function is to act as a 'switch' for the nearby gene; we'll call these regions switch DNA. The switch DNA acts as its name implies—it controls the nearby gene just as a light switch controls a light. How does it do this? Sticking to these switch DNAs are the gene-regulator proteins. The most important thing about gene regulators is that they act as signals to control the huge set of proteins—an astonishing 40 or so different ones—that are needed to make messenger RNA. If there is no gene regulator stuck to the switch DNA then, quite simply, there can be no messenger RNA made from the gene and therefore no protein. Controlling the activities of the gene regulators is the cell's main problem in regulating its protein production. From this viewpoint, the role of the gene regulator is the central player in the genetic drama that is unfolding and will become the key to understanding how the flow of information out of genes is controlled.

Geneticists frequently talk about genes being 'switched on' or 'switched off'. What we mean by this is that the information within a gene must be released within the right cell at the right time: we talk about a gene being 'switched on' in a hair cell to make the hair protein keratin. Equally, hair cells do not need to make blood proteins and so the genes encoding blood proteins are 'switched off'. In short, a gene can be 'on' or 'off' just as a light can be on or off and it is the gene regulator that controls whether the switch is on or off. A gene regulator turns the gene on by sticking to the appropriate switch DNA and turns it off by becoming detached; some gene regulators work the other way round, turning a gene off by sticking to the switch DNA and turning it on when they detach.

But there is a problem with this: if a gene regulator is the key to switch a gene on or off and we have 50 000 genes, we must have 50 000 gene regulators . Even worse, if we have 50 000 gene regulators, which are themselves proteins, they must be coded in 50 000 genes, which in turn must have 50 000 gene regulators. (This story is becoming ridiculous; we would appear to require gene regulators ad infinitum!) So the simple idea that each gene is controlled by its own gene regulator very obviously cannot be correct. How then does the cell control its 50 000 genes?

The exact number of gene regulators in human cells is not yet known but we can make a guess based on much more detailed understanding of the single-celled fungus we know as yeast, used to make the alcohol in beer and wine by fermentation. There are 6080 genes (give or take a few) in a yeast cell because the detailed order of rungs in the entire DNA of yeast has been established: of these, about 170 genes are probably coding for gene-regulator proteins. Humans probably have the same proportion of gene regulators to genes as yeast does (this estimate may not be completely accurate but will be close enough not to make a difference). If this assumption is correct, then humans must have about 2800 gene regulators. How can these control 50 000 genes?

Our understanding of how gene regulators operate in humans is based on a detailed knowledge of a relatively small number of genes, but this clearly shows that human genes are normally controlled by more than one gene regulator; in fact, as many as six at a time are needed. This completely alters our view of gene regulators, because if just two gene regulators control each gene then there are about 7.8 million (roughly 2800 times 2800) pair-wise combinations of regulators that could be used. But if all genes are controlled by six regulators, then there are about 570 million million million combinations for the cell to chose from! The large number of possible combinations of regulators and the (relatively) tiny numbers of genes means that the cell has all the possibility for control that it needs.

Does this mass of gene regulators explain all aspects of the life of cells? Far from it, because cells, as they divide, can change the genes they use and therefore alter their properties. This means that a cell has a history—rather limited, admittedly, but made up of how many times it has divided and what proteins it has made in the past. A cell seems to have some mechanism that enables it to keep a record of its past history as well as an ability to sense its current position in the body; if you combine these you have a reasonably good description of how cells can determine what proteins they should produce. This sounds as though it might be quite a complicated record but, fortunately, you would be wrong to believe this; the cell does not have to keep a neat central master record of its history, which can be consulted and read, because each gene carries its own record written in the language of gene-regulator proteins. What does this record look like and how might it work?

Cell historians and reading history

You will recall that gene regulators work in combinations: let us say the gene that makes a hair protein requires six different gene regulators (five would not be enough for the gene to be switched on). Now let's consider these six gene regulators, which I will number 1–6 for ease. Where did they all come from? Did all six have to be made in the cell once the cell 'realized' it was a hair cell and needed to make a hair protein? The answer to this question is that it is what has happened to the cell before it became a hair cell, its past history, that is a key factor in determining the final events that go to make a hair cell produce hair proteins.

The most distant ancestor of our hair cell was the fertilized egg: many cell divisions have occurred since then and the cell's ancestors probably made many different proteins and necessarily many different gene regulators. When a cell divides, the proteins it contains, including its gene regulators, are divided into the new daughter cells. This means that the cell can in principle keep track of what its genes were doing in the past by simply 'looking' at the gene regulators it contains in the present; the cell will have a 'record' of what genes its parent cell used. The handing down of gene regulators is, of course, necessary to allow the new cell to continue to make the appropriate proteins.

Now comes a very clever piece of control. We know that a single gene regulator can be involved in controlling several different genes and so it is quite possible that the ancestors of the hair cell made gene-regulator 1 to switch on appropriate genes. In these ancestor cells our hair-protein gene will remain switched off because only the first of the six gene regulators required is present in the cell. As the ancestor cells divided, drawing closer to the generation that includes our hair cell, then gene-regulators 2, 3, 4, and 5 are made so that they can switch on the necessary genes. The parent of our hair cell now contains all of the gene regulators except number 6 and its lack will keep the hair-protein gene switched off. It is only when the hair cell is finally formed by the division of its parent that gene-regulator 6 is made. Now all of the controllers are in place to switch on the gene and make hair protein. In short, the cell's history has led inexorably up to this point. All that needs to happen is that the cell makes gene-regulator 6 and it is inevitable that the hair-protein gene will switch on because that is what the history of the cell has set up all the gene regulators to do. This is very clever because it means that by the very act of switching on a gene regulator, a cell not only controls what proteins it can produce in the present, it also sets up the control of the proteins it will need in the future; in short, the act of control is not only for the moment, it is also for the future.

Much of the complex control of genes can be explained by the activities of the gene regulators, but this is not really sufficient to account for the wonderful way that cells act together to make up a human being. Even if we knew how all of our genes were controlled (which we most certainly do not!), we would still not understand the workings of our bodies because genes are simply a store of information. Proteins have the key role in the life of the cell and it is their inter-actions, each dependent on their exact chemical structure, that enable them to define the properties of cells; in turn, it is the interaction of whole cells that further makes our bodies. This should be an obvious comment but one that is sometimes misunderstood even by geneticists, some of whom seem to imply that the Human Genome Project will tell us how cells themselves function, which is a grossly misleading misrepresentation of the project. The incredible complexity of the properties and behaviour of cells lies within the levels of organization of protein interactions and protein activities and not solely within the types of protein they make.

It sounds as if there should be no simple rules to how these further levels of organization are achieved but, surprisingly, this is not the case because the body is organized using a very flexible 'toolkit' of actions; it is to these that I now turn.

■ Genes and the body plan: how to make a human

So far I have shown that genes store information; this is true for all living organ-isms—animals, plants, and even simple organisms like bacteria. Now we need to ask what it is that makes Jean and Jeanne have bodies that have a specifically human shape, a human 'body plan', unlike that of any other living creature. Once again, the answer is in genes, but this time it is by the way the cells of Jean and Jeanne's bodies control the release of the information stored in their genes to make proteins, and how these proteins are then used.

The development of a human being from an egg to an embryo is a stunningly complicated task and, for it to occur, every cell must successfully make many decisions as to what protein it should make or what its fate should be—to divide or not to divide, to live or to die, to move or to stay. How are all of these decisions made? In the most general of senses, is each decision a unique problem with a unique solution, or are there more general solutions that are used over and over again in different contexts and on different occasions?

From the point of view of geneticists, the worst case would be that every decision a cell makes is different and in consequence the whole problem of human development would be completely intractable to our understanding because we would have to understand literally thousands of different decisions in a trillion cells. Mercifully this is not the case and, instead, the organization of our bodies comes about using just a few underlying actions that are used over and over again—admittedly in different contexts and often involving different proteins, but nevertheless with a shared fundamental similarity.

The 'toolkit' of development contains just five main types of tool, which seems a rather paltry collection with which to build a human being. The important point is that just as you do not need a different kind of screwdriver for every screw you wish to undo or screw in, so the human tools may be used in many different contexts, giving virtually infinite flexibility. That this is true is perfectly demonstrated by the realization that all animals use the same toolkit to build themselves, with resulting shapes that are as diverse as a human, a fly, a worm, or a whale. Once again, I shall leave out discussing the huge mass of detail that is inevitably associated with such diverse and complex outcomes and focus on the general by looking at each tool in the toolkit in brief outline. What we will discover are five fundamental methods by which animals control their development:

- Cascades: how genes control themselves and turn on in order.
- Signalling: how cells communicate and sense their surroundings.
- Positioning: how a cell knows where it is in the body.
- Division: how cells chose to divide.
- Fates: upon demand, cells can move, change shape, or die.

I will discuss each of these in turn.

Cascades: genes in order

The simplest single idea behind the development of a human being is the use of gene regulators to switch on or off the appropriate genes at the appropriate time and place, as discussed above. This uncomplicated idea is fine, but it is really a little too simple because the properties of a cell are defined by the collective action of many different proteins with different functions. The challenge that faces a cell must therefore be to make unrelated proteins, in some co-ordinated fashion. How do gene regulators contribute to this goal? First, by generalization: one gene regulator can control more than one gene; and, second, by a carefully constructed hierarchy of occurrences, where one event leads inexorably to the next.

The first point is uncomplicated: if a gene regulator can contribute to switching on many genes, then this is an easy way for a cell to achieve its goal of making the relevant proteins at the right moment, simply by switching on the correct gene regulator at the appropriate time—just as happens for the hair-protein gene that I discussed earlier. But this is a rather static view of the life of a cell, which inevitably has a history that might well have involved making many different collections of proteins. To achieve this goal, the cell uses a hierarchy of gene regulators.

To illustrate what this means, let us think about four gene regulators, GR1, 2, 3, and 4. Let us also suppose that GR1 is present in the egg and its job is to switch on a group of genes, one of which includes the gene for GR2. GR2 in turn switches on many genes that include GR3 and similarly GR3 switches on, amongst others, GR4; and so on. At once this simple idea, which is compatible with everything we know about the roles of gene regulators, suggests how cells control sets of proteins that are needed in a fixed succession. This is a 'hierarchy' of gene regulators, so GR1 is at the head of the hierarchy, then GR2 then GR3, GR4, and so on throughout the life of the body; as each gene regulator is produced, so it can switch on its set of proteins, resulting in a 'cascade' of new proteins being made in the cell. The beauty of this idea is that an event must always precede its later consequences; GR2 must be made before GR3 can turn on its target genes, providing an elegant simplicity to the timing of events. Of course, the simplicity is deceptive: human cells do not have just one cascade in operation at any one time and at present we really have no idea how many are working in a cell at any specific moment.

This idea could be taken to mean that the cascade of gene control has a dynamic of its own; once the cascade is initiated, it must run inexorably to its finish. This view would be wrong because the decision of a cell to switch on the genes in the next step in the hierarchy is subject to control by the cell. In particular, the cell must have some ability to relate its own activities with those of its neighbours and, to achieve this, it must be able to sense its surroundings.

Signalling: partners and light switches

A cell cannot possibly be a simple isolated bag of proteins that is unable to communicate with the outside world: if it were, then there would be no way it could co-ordinate its activities with those of its neighbours and the human body would not exist. It follows that communication between adjacent and more distant cells must be a major requirement for the normal functioning of a cell and this is achieved by the properties of a whole group of proteins that I will call 'signalling' proteins.

These proteins are rather elongated molecules that are spread across the fatty membrane that surrounds all cells, so part of the protein is inside, part outside. The part on the outside of the cell has a shape that matches the shape of a second protein (its 'partner'); the partner sticks with a perfect fit to the signalling protein because they have complementary shapes, much as a lock and a key are complementary. The coming together of signalling protein and its partner is a key event for a cell because it is the start of a series of events that tells the cell that something is happening on its outside and to which the cell can respond. This information is transmitted to the interior of the cell as a series of simple chemical changes that are started by the interior part of the signalling protein, one protein modifying the next, in a wave of chemical alterations. Each protein modified in this way acquires new properties, and so this results in the co-ordinated acquisition of a whole set of new protein functions.

It is the gain of these new functions that means that an external partner protein can trigger a very specific response when it is detected; it might start producing a set of new proteins or stop production of another, start cell division or stop it, and so on. The specificity of interaction between the partner and signalling protein allows a cell to have many different signalling proteins on its surface at any one time and each will trigger some response in the cell. How they interrelate is best explained by analogy with the lighting in a house. A house has numerous light switches that control individual lights, but also several lights may be controlled by one light switch and several switches may control the same light. This is just as it is in a cell, where different partners can cause the same event (for example, cell division) or different events may be triggered by one partner (for example, divide and make muscle proteins).

A cell can 'sense' its surroundings via its signalling proteins and so an important question is the nature of the partners they can detect. In some cases the partners are attached to surroundings cells, so one cell will know it is in contact with another, or they may be freely floating in the liquids between cells, in which case the cell is detecting its liquid surroundings. In either event, the act of detecting the partners triggers events in the cell that are the necessary response to the information the cell is receiving. Of course, different cells contain many different proteins and so a refinement to this whole system is that the same signalling/partner pair can have a very different effect on different cells: there are numerous examples during development where the same partner protein is used to achieve very different goals.

The number of signalling/partner pairs in human cells is not known with certainty, but from the results of the Human Genome Project we know that one of the largest classes of proteins are signalling molecules, which must give our cells a highly sophisticated ability to determine their surroundings in the body. Sophisticated or not, however, it is not sufficient for a cell to know only who it is near; it must also be able to receive direct instructions as to where it is.

Position: the longitude and latitude of all animals

One of the most surprising and important discoveries made by geneticists in the twentieth century was the realization that there is a universal way of giving cells precise information as to where they are located in bodies as different as flies, worms, whales, or humans. Perhaps, with the benefit of hindsight, we should not have been so surprised at this because at a fundamental level all animals are rather similar. This last statement must seem very counterintuitive: what is similar about the body form of a fly, worm, whale, or human being? The answer is that we are paying too much attention to superficial features such as size or the presence of fins, wings, or legs and not enough to more fundamental features. In each animal, the head is at one end, the rear at the other, each has a front and back, and left and right sides are virtually identical.

Most animals can be described in this simple way and once you have identified these fundamental features, then it is less difficult to define position rules that place each animal's specific features on this fundamental plan and which define the location of the head, legs, arms, wings, chest, tail, and so on. For example, heads are at one extreme end; paired wings, arms, and legs are variously located between head and tail; and other features are distributed in the front-to-back dimension. These rules simply state where things should be and after that an animal can have any variation—wings, four legs or two arms and two legs, antenna, eyes on stalks, noses on legs, eight eyes or two, fur, spikes or scales: the variations are endless. So, in its simplest form, an animal's body can be described by features located in the head-to-rear and front-to-back directions.

Each of these directions is called an axis of the body and all animals define position by relative location along the two axes—just like an internal latitude and longitude. So the middle of your head is about a tenth of the way down the head/rear axis and about halfway down the front/back axis. Your navel is about halfway down the head/rear axis and at the front end of your front/back axis. In short, every feature of your body can be positioned by defining the position along the two axes.

The obvious question is how do cells achieve this: what is the cellular equivalent of a sextant? The answer is that the 'cellular sextant' is made up of whole groups of proteins—almost all gene regulators—whose function is explicitly to tell a cell where it is. To understand how the cellular sextant works, let's consider a simple organism that has only a head and a rear and nothing else; imagine it has no front/back thickness at all. Cells in the head are told they are in the head because they make a 'head' protein and cells in the rear make a 'rear' protein. What about the cells in the middle? Do they know they make a 'middle' protein? Not necessarily, because if they made both 'head' and 'rear' proteins, this would uniquely define them as 'middle'. This is indeed exactly how cells are told position, by the presence of a potentially overlapping set of position-defining proteins. In fact, in human beings, there are many, many proteins involved in this role; we have the equivalent not only of 'head' and 'rear' proteins but also proteins of 'middle' and many other intermediate locations.

The best understood of such proteins is a family of gene regulators called Hox proteins that were originally discovered in the fruit fly, an insect whose development is better understood than almost any other organism (geneticists have been studying fruit flies for many years). A series of very elegant and remarkable experiments have shown that the various Hox proteins are made in different regions of the fly's body, and so the cells in a fly 'know' their position by 'reading' the combination of Hox proteins they contain—a combination called the 'Hox code'. The really stunning discovery was that all animals, from the simplest worm to human beings, make Hox proteins and use the Hox code as their internal sextant.

This discovery was of huge importance because the fact that Hox proteins are gene regulators immediately suggests that the reason cells in the various parts of our body have distinct characteristics is because they contain different gene regulators and so make distinct proteins. By this view, reading the Hox code is not like reading the DNA code: it does not have to be decoded because the code is comprised of the gene regulators that will control the cell's interpretation of what the code means. A set of Hox proteins in a cell located at the midpoint of an animal will switch on the genes that will give rise to a leg in a human, but a wing in a fly.

The Hox code is just one example of the similarity of developmental processes among many animals and other gene regulators have been shown to have key roles. One of the most spectacular examples is the protein called PAX6, which in both fruit flies and humans is involved in controlling the development of the eye, even though our eyes and those of a fly are totally different in structure. In humans and flies, lack of PAX6 results in eyelessness.

Why should the cellular sextant in animals be so similar? The reasons are actually not so profound. Quite simply, making a system that is capable of defining head and rear, and front and back, is complicated. So once it had evolved in its most simple form perhaps a 1000 million years ago, there was no real question of inventing a completely new system every time a new body shape evolved; indeed, it would perhaps have been more surprising if anything so complex had been invented more than once! The evolution of the cellular sextant is one of adding additional protein members—flies have 9 Hox genes whereas humans have at least 38—as well as altering their patterns of expression.

This is why the same fundamental mechanism can make a whale, a worm, a fly, or a human, simply because all the system is doing is defining the two axes. Where we would expect differences is in the detail and this has turned out to be the case; making legs in flies is different from making legs in mice, even if the methods for defining where legs go relative to the axes are similar.

This may be clever biology, but all we have done, if you think about it, is push the problem back one step: somehow the cell has to be told to turn on the appropriate set of Hox genes in the first place. The way this is achieved is that cells are initially given information about their position from signalling proteins and partners, before the body axes have even formed. Cells that are distant from their signalling protein sometimes make partner proteins. Given the partner has to pass through the space between the tightly packed cells of the developing embryo, it is unsurprising that most of the partner is close to where it was produced and there is far less at a more distant point; we call this variation from maximal amount of protein to low amount a 'gradient' of amount. Such a gradient immediately can be used to define direction (up and down the gradient) and also position (the actual amount of the partner indicates how far away the cell is from its source).

This is exactly how the earliest initial axes of the body are established in the embryo; special groups of cells, called 'organizers', make partners whose gradients define the axes. Only when direction has been established do the far subtler and complex patterns of Hox gene products come into being.

I have focused on the idea that genes are the key to development. Clearly genes are of central importance to all of these systems because they have the ultimate responsibility of ensuring that the right proteins are present in the right cells. This centre of attention on genes, however, is dangerous because it provides a false sense of our understanding: the properties of the cells in our bodies are the product of the functions of the individual proteins they contain and not just of the simple list of proteins they make. It is a little like supposing that turning on an engine is all a racing driver has to do to win a 'race'; if the engine is not turned on, the 'race' cannot be won, but winning needs more than a working engine! So let us move on to consider some of the important decisions a cell can make once it receives its positional instructions and that result in winning the race to build a human being.

Divisions and sitting around doing nothing

All of these ingenious systems for signalling have a limited number of effects on cells. The fertilized egg has to grow from a single cell to about a trillion cells and obviously cells have to divide to achieve this growth. Because each cell division results in a doubling of cell number, we can work out that about 40 rounds of cell division would be required to make the necessary number of cells; therefore division of cells and its control must be a key action of development.

How cells divide has become quite well understood over the past 10 years, even though studying the behaviour of a single cell in a human body is very close to impossible: the breakthrough discovery was that the 'simple' single-celled yeast could be used to study cell division because, like development, the control of cell division is similar in all cellular organisms.

The life of all cells is slightly monotonous and you may be surprised that cell division in adult humans is actually relatively uncommon; most cells are simply sitting doing nothing, neither dividing nor dying—they are in a state of 'quiescence'. In contrast, cells in a developing embryo are dividing rapidly and very purposefully. Obviously cell division must require that a cell makes a new set of its own DNA and proteins, and much of its life is spent doing just that. Making a copy of DNA must be particularly difficult. How does a cell know to make exactly one copy of DNA? What stops it making one and a half or two copies, which would be disastrous? The answer is that multiple tasks have to be carried out in a co-ordinated sequence. Copying DNA might be an initial event, therefore, but the process has to be checked to ensure that re-copying does not occur; the copied DNA has to be 'read' for errors and only then divided into two so that it can be passed into the daughter cells. Each of these tasks, and more, are carried out by collections of proteins that are normally inactive in the cell but, once the decision to start division is made, become activated in sequence, very often by having simple chemical modifications made on their constituent proteins.

After activation, the protein collection carries out its task of copying DNA, or whatever, and afterwards the individual proteins are either destroyed or their modifications are removed so that they become inactive again. Controlling all of these events is complex and is achieved by master proteins that generate timing mechanisms through cycles of activity and inactivity. One group of such proteins, 'cyclins', comprises very unstable proteins; they are rapidly destroyed after they are made and so have a very short life in the cell. They also have the unusual property of being able to switch off their own gene and this enables them spontaneously to generate cycles of high and low amounts of cyclin protein. This is quite easily achieved. As the amount of cyclin in the cell increases it will stick to the switch DNA of its own gene and cause it to switch off. But the cyclin protein is unstable and so the protein will be rapidly destroyed and cyclin levels will fall, allowing the gene to switch back on, which sets off a new cycle of high and low levels.

The role of cyclins in the cell is to modify proteins that are carrying out the various tasks of the dividing cell. Given cyclins are cyclically present in a cell, the proteins they modify will similarly be modified in a cyclical fashion: the variation of cyclin levels is the ticking clock of the cell that times the events of cell division. Unsurprisingly, the actual series of events in cell division is more complicated than a single master protein controlling all events and there are several such proteins.

An important feature of these controls is that they are exercised on the cycle only at key points and not continuously; this is very important because the cell does not have inexorably to progress from one round of division to the next. At key points in the cycle, the cell stops and decides if it will divide and, if not, what it is required to do next. The way the cell makes these decisions is to use input from its signalling proteins, which control the activities of the master proteins that themselves control division. This neatly brings us back to the start of this section: the cell's decision to divide is ultimately controlled by its surroundings and this is why cell division can occur with such extraordinary rapidity and accuracy in the developing embryo. The only difficulty from our point of view is that this throws back the control of development onto the many signalling systems of the numerous cell types in our bodies and the countless surroundings that they occupy. It is unsurprising that despite a huge amount of research, we remain very ignorant indeed of the final picture and our ignorance is made more profound by the realization that cell division is not the sole activity of a cell; it is to their other fates that we need to move.

Fates: move, change shape, or die

It would be quite incorrect to give the impression that all that has to happen in development is for cells to divide. Quite obviously the cell has to be in the right place at the right time, so how do cells end up in the right place? Do they just naturally divide and end up, many divisions later, in the right region or do they move around? The answer is that both processes occur: cells passively arrive in the right place and also some cells actually move to the right place.

The greatest of the long-distance runners are cells from within the developing spinal cord. Quite late in development, these leave the spinal cord along well-defined routes and, migrating along the skin's surface, they lodge at some distant position and start to make a dark-brown protein called melanin. These cells, called 'melanocytes', protect us against the damage caused by intense sunlight because melanin is very good at absorbing ultraviolet light—the component of sunlight that causes sunburn. When (and if!) we go brown after being in the sun is entirely due to melanin being made by the cells in our skin that originated in the spinal cord of the embryo.

There is a nasty sting in this tale: these spinal cord cells are very susceptible to damage by sunlight, particularly if they make only small amounts of melanin—as is the case in most 'white' people. This damage causes them to become cancerous, which can be very dangerous. Although the cancer might be only a patch of cells on the skin's surface, in some forms of the disease (the melanomas) the cells seem to revert to their original behaviour and start to migrate around the body, spreading the cancer as they go. A tiny patch of damaged skin cells may, therefore, kill by inducing 'secondary' cancers throughout the body very quickly—a thought that should perhaps remain in the forefront of our minds as we lie on the beach soaking up the sun.

These melanocytes are unusual in the body; few cells migrate such distances but many cells move shorter distances. Surprisingly cell movement does not have to accompany the movement of cells: a strange comment to make but its meaning is well illustrated by the case of the spinal cord. The spinal cord in an adult is a hollow tube running the length of the body and mainly consisting of various types of nerve cells. In a very young embryo, the nerve cells are found in a flat plate of cells down the length of the body: gradually, over a period of days, the flat plate curls up at its edges and becomes a tube. In humans, this process sometimes fails to be completed before birth and the baby is born with a condition called spina bifida; if a large region of the spine is open, this may cause the baby to die or, if a smaller region is open, surgery may be needed to close the tube up.

How does the plate of nerve cells roll into a tube? It is not by movement of cells but by cells changing shape and forcing the plate to curve into a tube. Many tissues are formed in this way; the eye, for example, is formed by nerve cells that form a sphere that then gradually turns into a cup as one region of the sphere is forced inwards by the change of shape of cells.

Movement and shape are important, but how does the body make something like a finger on a hand? Does the finger develop as a group of cells growing out of the palm? No: all of the fingers develop from a flat, paddle-like hand and cells between the fingers have a very strange developmental fate; they are born to die. Conveniently cells located between the regions of the paddle that will become our fingers die, causing the fingers to emerge from the paddle much as a sculptor cuts away the stone to reveal fingers cut from solid rock.

This type of death is actually rather complicated and does not happen by accident: in a very carefully controlled process, special 'death proteins' are produced that deliberately destroy the cell. This process is called 'apoptosis', from the Greek word used by Homer to describe leaves that fall in the autumn. Cell death by apoptosis is very important in defining the shape of an embryo and consequently many of the features of our bodies.

A conundrum: what makes us different?

The shape of the human body, the human body plan, is ultimately defined entirely by our genes because they store the information needed to make the proteins that work together, through the toolkit of human development, to make the human form. It is not a direct responsibility; that lies with our proteins and their functions in the cell. This leaves us with a conundrum. If genes determine the human body plan, even indirectly, what is it that allows every human being to look different, such that we have no difficulty in recognizing each other in a crowd of thousands? It is this problem of difference that I address next.

■ Genes and differences: the giant and the midget

Jean and Jeanne are unique individuals with many distinguishing features and abilities that are in part the results of the different paths of their lives. This cannot obscure the fact that they also were born with differences between them, just as all human beings are. The question that we therefore need to discuss is if genes make Jean and Jeanne human beings, what makes each of them so different? The surprising answer is that genes not only make them human, they also make them unique human beings.

Human uniqueness is not an illusion: we really are different from every person who is living now, who lived in the past, or who will live in the future. This is something to wonder at. Can you imagine a machine that is different every time it is built and yet is always fundamentally similar? A car that is always the same model but always looks different? Perhaps we could have an entire range of Porsche Genetic 911s that could be guaranteed to be individually unique but still an obvious Porsche?

Thinking about uniqueness is not easy, precisely because what you have to think about happens only once and can never be repeated. Somehow the extraordinary set of plans contained in our DNA has to be capable of generating the human body form but also capable of generating a huge range of diversity. Humans can be as tall as 8 feet (2.4 metres) or as short as 3 feet (0.9 metres), as light as 45 pounds (20 kg) or as massive as 500 pounds (227 kg). We can have white, black, brown, or freckled skin; we can have red, black, brown, gold, or blond hair; and we can have brown, blue, green, or hazel eyes. Our muscles can be large or we can have a slight frame; we can be 70 per cent fat or as little as 5 per cent; and we can be healthy or sickly, clever or stupid, witty or dull, funny or sad. The very essence of a human is that he or she is different from every other human in the world, with a different history, experience, and life and yet we are all identically human beings.

So we are presented with a dilemma: how can similarity and dissimilarity be reconciled? The answer is perhaps the most difficult thing for a geneticist to explain because it is based on an interplay of the individual events in a person's life—their lifestyle—with their genes; but genes, surprisingly, are not quite as constant as we might have thought from their central role in the cell. So let us move on, first, to think a little about the different lives we all lead, and how these might influence how we look and behave, and then return to the unexpected influence of genes on our individuality.

Our lives are us?

It is very clear that physically we are profoundly affected by the lives we lead and this is reflected in some very obvious ways: a weight-builder increases muscle bulk by simply using the appropriate weights to exercise specific muscles; a runner trains to develop muscles to run fast; if we eat too much we become fatter. None of these observations is in the slightest bit surprising because we are all aware that we can influence many aspects of our lives by the choices of lifestyle we lead. Perhaps more surprising is the realization that these types of influence are present even in the womb, where a baby is receiving all of its nutrition through the placenta. The mother's lifestyle, eating habits, and health will naturally have a direct bearing on the quality of the baby's nutrition; in the extreme, if the mother is starving, the baby will develop very poorly and be small compared with its well-nourished potential.

Some startling new ideas about how our lives as embryos might influence our lives as adults have emerged during the past few years: there is increasing evidence that our birth weight has a significant effect on some of the diseases that may affect us later in life. In particular, low birth weight seems to be associated with an increased chance of developing some types of heart condition and diabetes much later in life. These studies seem to suggest the very intriguing possibility that the amount of nutrition a baby receives in the womb can alter the properties of cells in a way that persist into adult life. How this happens is not at all clear, but if this research, which is incomplete, turns out to be correct, then the causes of other disorders in adults may be similarly hidden in the life of the embryo. The clear message is that we should be very careful about what we assume is an influence on our bodies; simply because we become sick now, does not mean that the cause must necessarily be present at the same moment.

We need some word to describe the millions and millions of outside events that influence us and we use 'environment' to describe these collectively. The question we seek to answer is, if our environments are all different, does this explain our dilemma? Are all the differences between us accounted for by differences in our environment? This is an attractive idea but it is wrong. To understand why and how we can know this, we have to look deeper into genes and how they store the information needed to make our proteins.

The tarnished helix

The ladder of DNA in our cells is a huge molecule containing many millions of atoms. It would be very surprising if anything as large as this was immune to damage, and DNA, like any other part of our cells, can indeed be damaged. In the extreme, the cell will be killed because it no longer can use the stored information in the DNA, but damage does not have to be as severe as that; what if a rung became damaged so that instead of a rung made of AT, we now have a GC?

These changes are very slight—what is, after all, one rung out of 3000 million? But what if the change was to a rung that was in a gene and this altered the amino acid that was encoded? This would alter the protein that the gene produced by one amino acid and this apparently tiny change could have a very important effect on the protein's function: it might be catastrophic. Imagine that a baby starts to develop with one of its proteins lacking its normal function: the cells that needed this function would behave abnormally and as a consequence the baby's development could be damaged, perhaps severely. Of course it is equally possible that the change to the protein's activity would be only slight, in which case the baby might be ostensibly normal and perhaps only the colour of the eyes, or the shape of the mouth, or any one of many small details might be affected. Indeed, this is precisely why we have different colours in our skin, eyes, or hair. For example, there is a change of one amino acid in a signalling protein called MSHR that causes people to have red or blond hair as opposed to brown or black hair. These types of gene differences are extremely common and occur in all of us, most of the time influencing us in quite small ways. This is why I am proposing to call all of these rung changes 'differences' because they are common and often of minor effect.

The DNA ladder can also lose a rung; this has a far more serious effect on genes. The DNA code is read in groups of three rungs, so the loss of a single rung will not only change the code word it is lost from, it will also throw out of register all the DNA after the lost rung. This is best understood with an example containing just three code words: AGT TCA AAT A, etc. If the first G is lost, then the code slips and becomes ATT CAA ATA, and so on. The new code words are completely different from the old ones and the effect is that no normal protein can be made from the point of the rung loss onwards; this almost always means that the protein is made with completely the wrong shape to function normally.

Unhappily, both types of gene differences can have major effects and produce very unpleasant disorders, generally rare, which are the source of great suffering and death; these are the so-called genetic disorders, such as thalassaemia, cystic fibrosis, sickle-cell anaemia, Tay-Sachs disease. These disorders are caused by loss or malfunction of a gene that, although not critical for human development, nevertheless is required for normal child or adult existence. These gene changes are rare precisely because they can kill the individual before he or she has reproduced and so the gene difference is lost from the human population when the affected person dies.

Many gene differences are not simply caused by chemical attack of the rungs of DNA because the task of copying DNA during cell division is huge; the cell has to make a copy of 3000 million rungs. Unsurprisingly this is always accompanied by errors; the proteins that copy DNA are remarkably accurate but make on average one error in every 100 million or so rungs. This means that we all contain DNA differences. Because we have 3000 million rungs, every time our DNA is copied about 30 errors are introduced. It is important that we understand that differences in DNA rung order are as much a part of our lives as are similarities; we all have the same genes but that is not to say all of the genes are identical. In fact, the DNA of any two unrelated individuals contains about one difference in every 1300 rungs. These differences are present partly because of errors that must have occurred when the copies of the DNA in the sperm and egg were made and also because any two unrelated individuals have different parents, who must themselves have many DNA differences. I will discuss the origins of DNA differences in much greater detail in Chapter 4. For the present we need to focus on the consequences of the differences rather than their cause.

How do we know it's your genes?

We have seen that DNA differences cause alterations to proteins in our bodies. The resulting change in protein function must influence our bodies and no doubt contributes to many of the human variations that I have discussed; that seems clear enough. But differences in our environment also cause variation in our bodies. This leaves us with a difficult question: how can we tell if the differences between us are due to the effects of environment or to gene differences? This is perhaps the central problem that geneticists face not only in carrying out their experiments but also in explaining their work to non-specialists.

The way the effects of gene differences can be detected is remarkably elegant and relies on the discovery that some people in this world are genetically identical—identical twins. The realization that identical twins must have identical DNA comes from the way twins are formed. There are actually two forms of twins—identical and non-identical—each with very different origins.

Non-identical twins are simply the result of two fertilizations happening at the same time; by chance, the mother released two eggs that were quite separately fertilized by two sperms and this ultimately resulted in the mother carrying two babies to birth. The twins can be the same sex or different, and except for the fact they shared a womb together and so are the same age, they are fundamentally no different from two children born to the same parents at different times. Naturally the non-identical twins will contain DNA differences, which, just as for all of us, will have been given to them by their parents and by errors in copying DNA.

In contrast, identical twins have a very different origin. A single fertilized egg starts to divide just as it does to make a single embryo, but for reasons that are not yet understood, the embryo, when it has grown into a ball of cells, breaks into two. Remarkably, this seemingly catastrophic event is actually harmless: where there was one nascent embryo, there are now two and both will grow to become normal human babies. But there is one absolutely critical and unusual feature: because both babies came from the same egg and sperm, they have identical DNA differences. These events might seem unlikely but about three or four births out of every 1000 are identical twins.

Two individuals with identical DNA provide a unique opportunity for understanding the effects of gene differences on human lives. Identical twins have exactly the same DNA and so it follows that any feature that is defined by gene differences should be identical between them. In distinction, non-identical twins have a mix of their parents' DNA differences and should share some differences but not others. In fact, if we study many non-identical twins we see that on average they share only about half of their DNA differences. Why only half? Because there is a jumbling of genes that happens when the egg and sperm are formed, which means each time the same couple have a baby, the baby will inherit a mix of gene differences from their parents. On average, each baby will have half its gene differences in common with its brothers and sisters.

We do not need to go into the detail of how and why this mix happens. For the moment all we need to accept is that identical twins have identical DNA differences whereas non-identical twins have on average about half their changes in common. This phrase 'on average' needs careful thought and we need to remind ourselves that it does not mean that all non-identical twins share 50 per cent of their gene differences: at the extreme some may share all of their differences and some may share none, but most will share some value in between. To establish the 50 per cent average means that many non-identical twins have to be studied before the 'average' will emerge—a sharp contrast with identical twins, where every pair shares 100 per cent.

The way twins can be used to test if some human variation is based on the effects of gene differences or is the product of environment is conceptually quite straightforward. To focus our attention, we can consider an experiment to test if our height is 'genetically' or 'environmentally' determined. To do this we measure the heights of a large number of identical and non-identical twins and an equal number of completely unrelated individuals. If gene differences contribute to variation in height then we should get the following results:

- Identical twins will match in height because they always have the same DNA differences.
- Non-identical twins will match in height half the time because on average their DNA differences are the same half the time.
- Unrelated individuals will match only rarely because their DNA differences are derived from unrelated parents and so only rarely will they be the same in two individuals.

In contrast, if gene differences have no effect on height and it is only the effects of environment—nutrition is an obvious possibility—that causes variation, then both types of twins and the unrelated individuals should be equally dissimilar.

This type of experiment has been carried out to study a long list of human differences by simply studying pairs of twins and asking if any physical differences such as height, weight, and body shape or mental difference such as personality and intelligence are similar when the identical twins are studied, whether they are similar half of the time when non-identicals are examined, and only occasionally similar in unrelated individuals. Perhaps even more interestingly, such research can be extended to study why individuals are susceptible to common diseases such as diabetes, asthma, cardiac conditions, and mental disorders. Here the study has to be modified by first identifying twin pairs of both types, at least one of whom has the disorder, and only then to ask if the second twin is similar or dissimilar; the first step is, of course, necessary simply because most twins do not suffer from the condition in the first place.

Such an investigation might seem a perfect way to study gene differences, but the results are far from perfect. Understanding why this is so will become a central theme of the rest of the book. In the clear-cut case of gene differences being the sole cause of a disorder, one would expect that each pair of identical twins would be exactly similar. For example, if one examined pairs of identical twins who had schizophrenia, one would expect that if gene differences were the sole cause of the condition, then each identical twin should suffer similarly. The actual result is that if one identical twin has schizophrenia, there is about a 50 per cent chance the other will be affected; for non-identical twins the result is 20 per cent; and for unrelated individuals it is just 1 per cent—the frequency of the disorder in the human population as a whole. This is a curious result; it clearly shows that gene differences must be important in developing schizophrenia because of the 50 per cent and 20 per cent similarity rates in the twins—figures substantially greater than the 1 per cent figure for unrelated individuals. But the results surely should be 100 per cent, 50 per cent, and 1 per cent, if gene differences are the sole cause of the disorder.

Why is there this difference? Quite simply there's an environmental effect: some part of the lives of these individuals must be affecting the chance that they will develop schizophrenia. Exactly what part remains frustratingly illusive and we cannot even decide at the most basic of levels how lifestyle might exert its influence. Does it contribute to developing or preventing the disorder (the actual observations are perfectly compatible with either possibility)? Schizophrenia is a relatively rare condition and it is a reasonable assumption that gene differences contribute to causing the disorder and that environment is likely to be an equally important factor in causing or preventing the condition from developing. This is a very unsatisfactory state of knowledge and it becomes even more so when we look at more common disorders such as asthma or heart conditions that affect up to 30 per cent of us. Twin studies show a very significant contribution of gene differences but we cannot even be sure that they contribute to causing the disorder; it is plausible that they contribute to preventing it instead and that the environment is the major cause.

It would be easy to get bogged down in detail at this point. To avoid this we need to draw a general conclusion about what we think twin studies actually tell us, which is quite clear: that a disorder or condition is most often influenced by gene differences and by the environment, and only rarely do we observe a 'pure' effect of genes alone.

Identical genes or identical homes?

But even this conclusion is not quite as solid as we would like because there is a very simple conceptual difficulty at the very heart of twin studies: pairs of twins, by definition, must have the same parents and so must often be brought up in the same household, by the same parents, and at the same times. Therefore twins are more likely to grow up in the same environments than unrelated individuals. This is a real problem because it means that twins could be similar primarily because of the similarity of their respective environments. The difficulty is not overwhelming: pairs of both identical and non-identical twins would be expected to share similar environments and so the critical part of the analysis is to show that identical twins are more similar to each other than non-identical, just as we saw for schizophrenia. Nevertheless we have a serious difficulty.

How we overcome this problem is elegant: use twins who have not been brought up in the same homes. Study twins that have been adopted apart at birth, by different adoptive parents in different homes. Such twins separated at birth are not common but certainly do exist and great efforts have been made to find them—because they are exceptionally valuable resources for the analysis of gene differences. There are research records on some 40 000 twins worldwide and so it is unsurprising to find that some identical twins have indeed been separated at birth and adopted apart. The main difficulty with this approach is that because such twins are rare, it is unlikely that we will find rare conditions in them, so many studies are impossible: if a disorder affects one in a thousand, the chance of finding it in enough pairs of twins to allow these studies must be remote.

Chance is a confounding presence in virtually all scientific studies: chance in twin studies is the possibility that results can be explained simply by 'accident'. The meaning of accident in this case is quite specific and best illustrated by looking at the example of the incidence of diabetes among twins. Diabetes affects about 10 per cent of the human population in the developed world, and so if we select any individual at random there is a 10 per cent chance that he or she has the disease: this is the key observation that will cause 'accidents'. Our concern is that if we compare just one pair of identical twins and see that they are both diabetic, we are unable to interpret the similarity. Are they similar because they have identical DNA or is it because 10 per cent of all individuals, either related or unrelated, will be diabetic? The difference is critical because it means in the first case that gene differences cause diabetes but in the second case they do not.

The problem is that using just a pair of identical twins can never yield a useful result because we cannot distinguish between the results of 'genetic' and 'accident'. But we can resolve this by looking at more twin pairs. The 'accident' comes about because of the 10 per cent chance, so if we use two pairs of twins the chance of both being 'accidentally' diabetic is now 10 per cent of 10 per cent, which is 1 per cent. As we use more and more twins, each additional pair reduces the chance of the 'accident' by 10 per cent. At some stage we can accept that the 'accident' is so unlikely that we can ignore it and, if this is so, then similarity must have a genetic origin. It is really only when we have analysed perhaps a hundred identical twin pairs that we can be sure we are not being fooled by the chance of an accident. This presupposes, of course, that we have a large number of twins affected by a disorder, which is frequently not the case.

Twin studies remain a powerful and slightly controversial tool of gene scientists with much of the controversy centred on their use in the study of intelligence and behaviour—a topic I will turn to in Chapter 8. For now I'll look at another problem: twin studies simply indicate an effect of gene differences, but how can we move from this most general of conclusions to actually finding out how many genes are different and what is their identity. For help with this, let's bring in the gene hunters.

Chasing genes

Finding the actual gene differences that are responsible for, or at least contribute to, the cause of human conditions and disorders is an enormous enterprise that requires luck, persistence, and a great deal of hard work. Fortunately even though some 10 000 conditions in human beings are influenced by gene differences, there is one general way of identifying which gene is actually involved in any one case.

The fundamental observation here is in principle simple: if a gene difference contributes to a condition, then it will always be found in individuals with the condition and not in those without. (In science, though, being 'simple' is not always the same as being 'true' and I discuss some very serious problems with this principle in the next section.) We have to keep in mind here that we have two of every gene, one from our mother and one from our father. At a simple level, therefore, we can expect both genes to be identically different in the individuals with the condition and perhaps only one in some individuals without the condition but who are, in the jargon, 'carriers' of the disorder (that is, individuals who, while not suffering themselves, could pass the disorder to their children). So the principle of finding the different gene is simple: just look at all the genes and find the one that is always different in the individuals with the condition.

It is, however, much too big a problem identifying the gene difference in this way—even with the information coming from the Human Genome Project—and so geneticists have developed a more indirect route. This relies on unusual regions of DNA that frequently are of different lengths in different individuals. Such DNAs—I will call them 'variable DNAs'—are not quite as peculiar as one might imagine because they are most commonly located in the DNA that lies between genes. Many such regions have been identified in human DNA and their rough locations have been worked out from experiments that have resulted in 'maps' of variable DNAs. These maps are dull and simple compared with maps of countries. For a start, they are linear and simply define the order of variable DNAs down the DNA strand, variable region 1 (V1) coming before V2, then V3, and so on to a numbing V10 000 or so!

Scientists can use this map to identify the actual gene difference involved in a condition by looking at variable DNAs rather than genes. How these experiments are carried out requires one of the most mathematically complicated analyses in genetics and it is unnecessary to understand this nor the detailed biological intricacies. The principle behind the experiments is that it should be possible to locate the V-regions that must be closest to the gene difference rather than directly detect the gene difference. If we analyse the V-regions of many individuals chosen at random then we will discover that they have many different V-region lengths. But if we analyse the V-regions of many people who all suffer from the same genetic disorder, then the V-regions around the gene are more likely to be the same, simply because they must all share the same gene difference. The sort of results that we expect is to discover that in our group of selected individuals there is a length of DNA defined by several V-regions that are far more similar than would be expected from analysis of people without the disorder: this must be the region that contains the disorder-causing gene difference.

In the jargon of genetics, a V-region close to a gene is described as being 'linked' to the gene, and so these studies are often called 'linkage analysis'. Two different ways of doing this involve either analysing families that have a disorder gene difference that is being passed down the generations or identifying pairs of brothers or sisters, or a brother and a sister, who are both affected by the same disorder or condition: this is the relationship of siblings (or 'sibs' for short), and so this sampling is often called 'affected sib pairs'.

The key achievement of such analyses is to pinpoint the region of DNA that must contain the gene difference and so define the region of genes that must be studied in detail until the final goal is achieved—a gene difference in all people with the disorder. The amount of work involved is startling, expensive, repetitive, and very prone to error and to false associations of V-regions with disorders, but it is also very successful. Scientists have identified well over 100 genes involved in human genetic disorders that are caused by differences within single genes.

Would that life were so simple

The discussion so far, though, has one enormous flaw that results in the gene hunter's work being far more difficult than I have suggested. The key problem is in the innocuous and apparently unexceptional comment that a causative gene difference '… will always be found in individuals with the condition and not those without'. This statement may not be true. The reality is that the effect of a gene difference can be masked by a favourable environment or by the influence of differences within other genes that mask or ameliorate the symptoms, resulting in the causative change being found in what appears to be a normal person. This is the gene hunter's nightmare because the feature of the gene difference that we were going to use to pinpoint the gene was that it was always found in people with the disorder and not in people without. The only way round this difficulty is to show by further studies that the difference is 'more commonly' found in individuals with the disorder than in those without. The nasty question (which I will explore in Chapter 3) is that we have no simple way of saying what 'more commonly' means; the result is that sometimes the gene hunter can identify a gene only as being 'likely', but not 'certainly', to be involved in a disorder.

Worse is to come; for many reasons (again, for discussion in Chapter 3) common genetic disorders are often caused by differences within more than one gene, which compounds the gene hunter's problems. The reason is simple: if differences within both genes have to be present to develop the disorder, then some normal people will have just one of the two changes and, of course, be unaffected. So, again, causative differences can be found in normal people and we loose the certainty of a one-to-one correspondence of gene difference with the disorder.

Does this mean that all gene hunting is hopeless? Of course not—I've discussed the success with the single-gene disorders—but it's hard and results have to be treated with caution. Initially the hunters will show that a particular gene is likely to be involved, but 'likely' is not the same as 'certain' until that likelihood becomes very large. In short, gene hunters can get it wrong, not because they have made mistakes but because there is no simple route to certainty: chance and certainty are the twin faces of genetics and we must not let the excitement of the latter overwhelm the inconvenience of the former.

Confounded chance!

As if chance and certainty were not enough of a problem, the two-step nature of hunting genes is the source of even more confusion amongst outside observers of gene science. This is because the first step—locating the region of DNA that contains a gene difference—is often accomplished and then released to the press with much publicity. Perhaps the best examples of this have been in the complicated fields of human differences, such as sexual preference ('the gay gene') or intelligence ('gene for genius'). The distinction between identifying the region and the actual gene is absolutely critical. Locating genes for disorders and conditions that are caused by more than one gene difference is an extraordinarily complicated and difficult task—as will become clear as the book proceeds; indeed, it may never be achievable. One of the reasons for this is that the whole analysis can be easily confounded by chance associations of V-regions and disorder, and these have given rise, and will continue to give rise, to false reports of gene locations; false, not because the scientists have made errors but false because that is the nature of the experiment.

It is perhaps hard to understand that the true nature of science is not the inexorable accumulation of fact and correct observation but is actually a process of suggestion, confirmation, or refutation, and of progress through both success and error. Perhaps nowhere is this more apparent than in linkage experiments. How can one sort out error from fact? With great difficulty, unfortunately, but I have developed some aides that are not rigorous, and are not really based on scientific reasoning, but help guide me through whether a scientific announcement in this area is likely to be correct:

- Do not believe scientists who announce their results to the media before they publish in the scientific journals.
- Attach less weight to reports of the identification of a region than identification of an actual gene difference.
- Always ask if other scientists working with other samples (a process known as replication) have confirmed the result.

Why are these three ad hoc rules useful? The first is because fellow scientists read scientific papers before publication and review them for scientific rigour and accuracy; then they make comments back to the authors—a vital process called 'peer review'. This means a published paper not only convinces the authors (why would it not, because they have a vested interest in publishing it?), it has also convinced two or three independent scientists. The second rule comes about because confounding errors are more likely when a large region containing a gene difference has been identified (once it is down to just one gene, the effect of chance is much reduced, although unfortunately not to zero). The third is perhaps most important; if another group of scientists independently can confirm the result, this adds weight to the findings. But beware: two wrongs still do not make a right!

■ Why one head and two arms and not two heads and one arm?

There is a very clear message in what I have discussed so far: that genes determine the form of our body but genes also contain differences that are one of the sources of human variation. These two roles suggest an obvious question: why is there no variation in the results of development? Why do we all have one head and two arms and not two heads and one arm?

The answer is not actually that mysterious or complicated: it is because genes work in a cascade where every event in development is defined by many different protein functions that collectively precede and trigger the next event. The important point is that collections of protein control development, not single proteins, and this makes the whole system relatively immune to change. Gene differences could indeed alter the function of one, or a few, proteins in the hierarchy but this will have only a very small effect on the whole structure, which is defined by the collective function of all of its components.

In contrast, it would be extraordinarily difficult to change the whole hierarchy to, say, making our head grow in the position of our leg. We can imagine that this might be achieved by reorganizing the entire hierarchy—by swapping a gene regulator at the bottom of the hierarchy with the gene regulator at the top. This would require an impossibly unlikely series of changes: genes and their regulators would have to be reorganized completely so they contained different switch DNAs and responded to the new gene regulators that had to control them; all the intermediate steps would have to be similarly redesigned, and, finally, all of the other components of the cell would have to be reorganized to fit with the new instructions and new requirements of their novel location. This is simply too complicated a series of changes to achieve and it is for this reason that humans develop with an identical body plan—one head and two arms and not two heads and one arm.

Gene differences probably have substantial impacts on human development because a high proportion, if not the majority, of fertilized human eggs fail to develop past a few weeks or months in the womb; this perhaps suggests that the programme of development is delicately poised on the edge of what is possible. A second piece of evidence points to a similar conclusion: an extraordinarily high proportion of human babies, about two in every hundred, are born with some clear sign of a problem in development. These defects can range from the minor (slightly out-of-focus vision caused by an imperfectly spherical eye) to the serious (a clubfoot or cleft palate), to the catastrophic (a baby born with one eye in the centre of the head or no head at all). Maybe there is some limit to the complexity of what genes can do, given their intrinsic differences; perhaps it is close to the development of the human body?

I started this chapter with three statements that I contended were all that we needed to know to understand the principles underpinning the roles of genes in our lives. I repeat them:

- Genes store information: they contain the information to make proteins, which are the chemical building blocks of our bodies.
- Genes and the body plan: they define the architecture of the body.
- Genes and differences: all human beings have the same set of genes but they can be subtly different in each one of us, making us unique.

The first point has most recently generated the greatest excitement with the approaching completion of the Human Genome Project, which I introduced early in the chapter. In reality this is the area that is perhaps the least important for us to understand in any detailed way. It is sufficient to realize that geneticists can read the code in our genes and that they have developed remarkably powerful abilities to manipulate and rearrange gene functions.

The second point contains the most unimaginably complex areas of our own biology—the development of a human being from embryo to adulthood to death. We can ignore most of this detail but grasp one critical point: genes make us human, with a rigidly defined body plan that is fundamentally immune to outside influence. This is the one area where genetic determinism is complete, overwhelming, and absolute—the iron fist of our genes.

The third point contains the essence of what it is that makes us individuals: the complicated interplay of genes and gene differences with our environment that makes impossible any crude genetic determinism of the forms and abilities of our bodies—the feather's touch of our genes.

These are the principles underlying gene science. Now we can explore how genes influence us, starting with the area that we hold most sensitively as being the essence of our humanity: how we differ one from the other.

A hundred thousand reasons
for being different

Even if Jeanne and Jean had possessed identical gene differences, the circumstances of their existence are so different that it is impossible to expect an equal outcome to both of their lives. But, of course, both Jean and Jeanne are not identical genetically and many DNA differences distinguish them; their cumulative effects will also contribute to their different future histories. These two influences are the reason why the life of one is a dream and that of the other a nightmare.

The reasons that the environment of the two Jeans is so dissimilar are almost entirely within the realms of politics and economics—topics that are not presently within the scope of our discussions. But how should we view the DNA differences that distinguish them? Are they rare differences, akin perhaps to the differences that cause genetic disorders found only in a relatively few families and not in most other individuals? Unexpectedly, perhaps, this is not the case: the DNA differences that separate the two Jeans are common and found in many, many other human beings; they are, in fact, very common indeed.

To understand why DNA differences can be so common we have to turn in a surprising direction: to the distant history of the human species and by appreciating this we will gain not only an insight into the differences between each and every one of us, we will also discover why such differences are an essential requirement of our survival and a primary contributor to the common causes of death. This is a curious and paradoxical collection of statements and understanding them will open up one of the most important areas of current genetics.

Death and gene differences

The idea that gene differences can be very common is of tremendous importance because about 80 per cent of us will die in part because of the effects of common gene differences. Here I mean the diseases and conditions that will kill most of us. The big 10 killers in the developed world are heart disease, cancer, stroke, lung conditions, accidents, pneumonia, diabetes, the AIDS virus, suicide, and liver diseases. Eight out of every 10 of us will die from one of these causes, all of which, except perhaps accidents, are in part caused by gene differences. Even suicides are affected by gene differences, because genes can contribute, as we have seen, to disorders such as depression and schizophrenia that often are a contributing factor in cases of suicide. Understanding these common differences is a key to understanding large areas of human disease and suffering, and so this is perhaps the most researched area of modern genetics.

The difficulty of understanding the idea that gene differences can be common really stems from our impression of the effects of single-gene differences on human life. At a last count, there were some 10 000 identified disorders in humans that are caused by differences in single genes and most of these are extremely rare; perhaps fewer than a few dozen cases have been recorded. To most of us this would suggest that all disorders and diseases caused by gene differences must be rare simply because gene differences are uncommon in humans. To understand how we can get out of this difficulty, we first have to discuss the basis of the common effects of gene differences on human beings.

Genes together

In practice, there are relatively few features of human beings that are defined by a single-gene difference. Almost always we see a continuous range of differences; for example, we see a few people who are very short, a few who are very tall, and the majority somewhere in between. Why? Because many genes act together to define height and so many possible gene differences can influence the outcome. If we had, say, 100 genes that influence height (the number is not known in detail) then differences in any one of these genes might tend to make you shorter or taller, but the effect would be lost in the sea of effect of the remaining 99 genes. If a person had changes in 10 of the 100 height-forming genes, all of which made him or her 'taller', then the chances are that they would be taller than average and this is, indeed, how genes contribute to height.

There is an important general point emerging here. Similar changes in many genes influence most features of the human body: weight, bone size, muscle bulk, fat content, heart-beat rate, oxygen consumption, lung volume, or, indeed, almost any other physical feature of our bodies. The intriguing question posed by this observation is if by chance you inherit all of the gene differences that contribute to, say, height, do you become extremely tall?

Let's consider the very intriguing possibility that by chance one person inherits all of the differences in the 100 genes that influence height, all of which would contribute to making the person taller. Presumably this would result in an unusually tall person. But what would happen if they were 4 metres tall—taller than anyone has ever been? The human body could not survive being this high; the bones or muscles could not support the body nor could the heart pump blood to the head over such a height. .

This would seem to be an unlikely occurrence. More probably somebody would inherit 90 of the hundred changes—perhaps they would only be 3 metres tall; but, again, this would be very difficult for the body to sustain. As we look for more people with fewer of the changes, we will find more and more and once we get to people of average height, we would find many people with very few changes. The general picture that emerges is that human beings range in height from the small number of people at the extremes, who have many gene changes, to the average, who are very numerous and have few changes.

The observation that a small number of people are at the extremes of human physical characteristics is an important point because such people frequently have medical problems; the classic example is that of people who are very overweight—obese in technical terms. Extreme obesity, suffered by men and women who weigh over 400 pounds (more than 180 kg), generally results in death at an early age from heart failure, but victims also suffer from many other problems including diabetes. In every respect, these people, at the extremes of the size range, are ill—they suffer from a disease called obesity—yet all we are looking at is simply an unfortunate individual who, by chance, is at one end of the normal spectrum of weights. The inescapable conclusion is that the extremes of the range for many human conditions are the same as having an illness. Viewed another way, 'illness' is inextricably linked to 'normality'—it is simply the far end of the spectrum compared with the middle.

This is a significant medical problem, not just the cause of a few rare conditions, and it embraces most of the commonest disorders that affect human beings, including asthma (which affects up to 30 per cent of the population in the developed world), various forms of diabetes (about 10 per cent), and migraine—overwhelming headaches that can affect the sight (4 per cent of children, 6 per cent of men, and 18 per cent of women). These are all very serious conditions—diabetes is the cause of 3 per cent of all deaths—but the fact they are so common is quite difficult, at first sight, to understand. The extremes seem remarkably common and why should asthma sufferers (30 per cent) be so much commoner than diabetics (10 per cent), or a few per cent of obese individuals. The simplest explanation for this, of course, is that environment must also play a part in these sorts of conditions. But we still have to face the question of why gene differences can be so common.

Two rares make a very rare, not a common

If gene differences are rare—as shown by the rarity of the 9000 disorders caused by differences in single genes—then how on earth can disorders caused by differences in many genes be more common? Disorders caused by single-gene differences occur in about 1 person in every 10 000. How often would we expect to see an individual with two disorders caused by differences in two genes? We can work this out simply. If both disorders occur in 1 in 10 000 people, then if we multiply the two together—1 in 10 000 times 1 in 10 000—we get the answer, a staggeringly rare one person in 100 000 000. Given there are about 6000 million people on our planet, we would expect about 60 poor individuals to have two disorders at the same time. Of course the possibility of having three, four, or five gene differences is even more tiny; we would never see them because there are not enough human beings for it to happen! So here's the problem: how can differences in several genes (perhaps five) in asthma, for example, cause a disorder that is found in 3 out of 10 people? It simply is impossible; the numbers do not add up!

How do we get out of this difficulty? The answer is unexpected and the discovery of the answer is truly revolutionary and it has opened a great chance of altering how we human beings will live in the future. It is also a very simple answer—differences within genes must be very common and not very rare!

Rare differences and common differences—common disorders and common causes

Gene differences are common! Is this a reasonable view? If gene differences are common then, of course, our 1 in 10 000 times 1 in 10 000 sum is completely wrong. Let us say all gene differences occur in one person in two: then the chance of having two gene differences is 1 in 2 times 1 in 2, which is 1 in 4; the chance of three gene differences is 1 in 2 times 1 in 2 times 1 in 2, which is just 1 in 8; similarly the chance of four gene differences is 1 in 16, and the chance of five is 1 in 64. In short, the numbers look quite convincing; they do not quite add up—five gene differences is still 1 person in 64, which is much less than the 1 in 3 of us who will die of heart diseases. I will come back to this in a while but, for the present, it is reasonable to say that if gene differences are common, then relatively common conditions can be caused by gene differences.

But hold on! If gene differences are so common, how can all the disorders caused by differences in single genes be so rare? Surely we have hit another dilemma? Not quite: there is a good reason why these single-gene disorders are rare, which I mentioned in Chapter 2: the symptoms are so severe that often they kill the unfortunate sufferers at a young age before they have reproduced and the gene difference will 'die' with them; most gene differences that cause severe symptoms can never become common.

This gets us out of our dilemma: differences in single genes that cause rare disorders often kill the person who has the gene change but it must also follow from this that common gene differences must not have severe effects. Does this make sense? Surely if common gene differences act together to make a serious disease, why are they not lost from the world like the rare differences that cause severe disease? Is this another impasse? No, we have two ways out: first, the gene differences individually do not have to have any good or bad effect and, second, the gene differences individually might be advantageous.

Let us think about these two points by considering how our five gene differences can act together to contribute to a disorder using diabetes as the example. Diabetes is an extremely complicated disease that occurs in many different forms but with a clear underlying problem—the body cannot control the amount of sugar it contains. This is very damaging because sugar is a vital energy source in the cell—energy used to make most of the chemicals a cell requires—and failure of control results in too much or too little sugar being available in the right place at the right time, which causes a whole host of unpleasant symptoms including coma, blindness, and death.

We do not know how many proteins are involved in controlling the amount of sugar in a body—probably more than 10 and fewer than a few hundred different ones, and this is why the disease is so complicated. In principle, any one of these proteins could have a difference within it that results in a breakdown of some part of the whole control system. Let us think about this in a bit more detail. If just one protein is different, the effect on the whole could be very slight; if two proteins are different, or three or four, the effects could be similarly small. But at some stage, perhaps when five or ten proteins are different, then the cumulative effect of all of the differences crosses some 'threshold' and the differences to the control become so obvious that the symptoms of diabetes result. Under this view, a person might have nine proteins that are different and no diabetes, but a person with all 10 will suffer from the condition. This means that individually it does not matter if a protein is different or not, because it has negligible effect by itself, and this is why gene differences that contribute to common diseases can themselves be common—they will have a bad effect only in those individuals that contain 10 of them at the same time.

Why do these gene differences not get lost from humans because of the death of people with diabetes? One simple answer is that some forms of diabetes only occur late in life—after childbirth—and so are not lost by the normal process. A more-intriguing idea is that these differences have a beneficial effect and so having some of them, but not all, can be an advantage. This may well be the case for diabetes: perhaps if you have five of the 10 proteins that are different your body is better able to absorb sugar. This could be of real advantage to you if you lived (as we all did 10 000 years ago) through periods of feast and famine—you might survive because your body is more efficient at utilizing the sugar recovered from limited amounts of food than was someone who did not have the proteins that are different. We can only speculate that this is the case for diabetes because we do not know the identity of the proteins that are involved, but there is good evidence that proteins that are different can be advantageous as well as disadvantageous.

So, it is very reasonable that common differences within genes must underlie the common disorders. Discussing this has been a long process—every apparently contradictory observation has to be understood and explained—but we are finally at the end of it. Except, inevitably, for one tiny problem: I have not explained why differences in some genes are so common in the first place, whether they are advantageous or not; this is now the key question that we must understand.

The special cases of being different

I have already touched on one reason why DNA differences can be very common; this is the special case where there is an advantage conferred on the individual who is different. Two good examples explain some of the reasons why northern Europeans predominantly have pale skin and why some people are relatively resistant to the ravages of malaria.

Sunlight is required to make vitamin D in all humans; cells can use the radiation energy of the ultraviolet (UV) part of the spectrum to power the necessary chemical reactions. The problem with this is that UV is also very damaging to DNA, as I have already mentioned (Chapter 2), and so there is a balancing act to perform: too little UV makes no vitamin D, too much causes DNA damage. People who originate from the tropics have gene differences that result in relatively dark skin pigments and these block the UV rays that cause the worst damage; some UV still passes through the pigment layer and makes the vitamin underneath the skin surface. In contrast, people native to northern Europe have very fair skins because there is so little sunlight and every bit must pass through to make vitamin D; they are rarely exposed to so much sun that their skin is damaged. This means that the gene differences required to make pale skins become more and more common the further north you pass in the world.

Malaria is a disease that affects 300–500 million humans every year and of which between 1 and 3 million will die, especially children under five. Its cause is a tiny organism called *Plasmodium* carried by a mosquito 'host' that infects humans through the mosquito's bite. The *Plasmodium* then grows in the red blood cells in the human victim and can cause some very unpleasant fevers that may result in death. Red blood cells are full of a protein called globin, which is part of the oxygen-carrying system. Remarkably differences within the globin genes result in globin proteins with properties that make red blood cells a much less attractive home for *Plasmodium*, and so people with one globin gene difference are made partially immune to malaria: if they are unfortunate to have differences in both genes they suffer from the common genetic disorder of sickle-cell anaemia. Because malaria is confined to specific parts of the world, the gene differences are also common in the same parts of the world.

In both these cases the gene differences became common in individuals from the appropriate regions because of the 'advantage' that they confer on the individuals carrying the differences. It is important to understand that 'advantage' has a specific meaning in this context and one that requires a rather different perspective on our lives—one that many people may find slightly difficult to accept. If we can forget about human beings for a second, all living things share one feature in common—the ability to reproduce more of their own kind. Looked at from the perspective of simple living creatures, such as the single-celled amoeba or a bacterium, the essence of life is to survive and to reproduce. I think this is a perfectly reasonable view of these organisms that are very ancient in origin and far less complex than human beings.

Bacteria and amoebae, like all living things, contain DNA and genes and, equally naturally, differences occur in their genes that can alter the way they behave or respond to their surroundings. So let us say a change occurs in a bacterium that alters the way it moves—say it can move faster (many bacteria move using a whip-like structure that they thrash around and which pushes them through the liquid-covered surfaces of their preferred habitats). If a bacterium can move faster than its neighbours then several things may happen; it might escape from danger by swimming and thrashing away from a bacteria-eating worm (there are many such bacteria-eating predators) or it might move faster towards sources of food than its slower-moving companions. In either event, the speedy bacterium is likely to have a greater chance of survival that its slow-whipped friends, and if it survives, or survives longer than they do, it is likely to have more time to produce offspring.

This means that if we were to look at the bacteria population over many generations (and as some bacteria can reproduce themselves in just 20 minutes, this is an easy experiment to do for real in a laboratory), what we would find is that, quite rapidly, the faster-moving bacterium's offspring would become more and more common in the population. The reason this is happening is because in each generation the fast bacteria have a greater chance of making more offspring than their slower-moving colleagues and so the next generation will contain a greater proportion of fast bacteria, whose offspring will, in turn, have a greater number of offspring, and so on. What we see is that the fast gene differences have given the bacteria an advantage in the competition to reproduce. The advantage can be quite small, but over a few tens or hundreds of generations, depending on exactly how advantageous the change is, the advantage is multiplied up until the fast bacteria are very common in all populations; indeed, some populations may exclusively be all fast.

From this simple example of what advantage is to bacteria, we can look at humans and realize that advantage must mean exactly the same. A gene difference that makes a human being more successful in surviving and reproducing, even if the effect is slight, will gradually become more and more common in the human population. It is perhaps a difficult idea to accept, but most certainly at one basic level humans are simply complicated devices that reproduce themselves. I do not wish to argue over the philosophical implications of this rather bleak picture of our existence but I do want to point out that this sort of advantage to gene differences is, of course, how some of them get to be very common indeed—blue eyes, globin differences, hair colour, skin complexion, resistance to diseases, all the underlying differences are common because of the advantage that was, many generations ago, caused by being different.

Does this mean that all common gene differences are there because they are advantageous? One obvious problem is that if all the differences within genes are advantageous, then how come we do not all have the same differences? Why do some of us not have the advantage of the differences? One solution to this problem is that the change could have happened relatively recently and so perhaps too few generations have passed for all of us to have the same, advantageous, change. This is, of course, possible, but we cannot be sure until we know much more about the effects the gene differences may have on us. Another way that DNA differences can become very common in populations, however, has nothing whatsoever to do with advantage. To understand this possibility we have to take a very unexpected route: we have to understand the origins of human beings and the history of the past 100 000 years.

Where do all the differences come from?

Jean and Jeanne are genetically distinct and their histories make this obvious, but what is not so apparent is whether they are unusual in this respect. Are all people different or just those that have widely different origins—say Africans compared with Europeans or Chinese compared with Indians? The reality is that gene differences are actually very common so that we are all different, one from another. To understand why this is so, we have to go back a step in our thinking about differences in genes.

In Chapter 2, I showed that about one rung in 1300 would be different between any two unrelated people: many of these differences would be in the rungs of DNA between genes because differences in genes tend to be slightly rarer, but this still means that there would 3 million differences in all. Are these differences uniquely found in the two individuals we are studying or are some of them found in other people? There isn't the experimental information to answer this question accurately. Our knowledge so far is that many of the differences would be found only in the two individuals but that as many as half of them—that is, perhaps 1–1.5 million (we cannot, as you will see, be more precise)—would be present in varying proportions of the people of the world: these are the 'common' differences that we have been discussing. What do these differences look like? Quite ordinary: for example, one piece of DNA could have the chemical composition of GATTCGA in half of the human beings in the world whereas the other half could have GATACGA. The difference is tiny—just a T instead of an A. What we have to explain is how half of us have the first and half of us the second rung sequence.

■ Out of Africa

To solve the problem of why we have so many common differences we have to go back in time about 100 000 years. Then, there were probably very few human beings. The distant ancestors of our species probably first appeared in Africa as long ago as 5 or even 6 million years ago, but people looking much like we do today—so-called 'anatomically modern' humans—were not present until about 200 000–100 000 years ago. We have limited information about how these early humans lived: probably they were organized into small groups that survived through hunting and gathering and who roamed at will through Africa. Several lines of evidence suggest that these first fully modern humans were successful in surviving in this way, and, as a result, bands of them spread out over the whole of Africa, and then into the Middle East and on through the Indian sub-continent to the East.

About 40 000 years ago (possibly 60 000 years ago), early modern humans crossed the sea, hopping from island to island—probably using some sort of simple craft—from Southeast Asia to New Guinea and Australia, and occupied the huge continent of Australia. At about the same time, the first modern humans spread into Europe. This process was completed only relatively recently—perhaps only 12 000 or 10 000 years ago—because until then much of northern Europe was covered by ice and in the grip of an ice age. At least by 14 000 years ago (perhaps earlier), people from central Asia had crossed the land bridge that joined Asia to the Americas (the land now under the cold waters of the Bering Strait) and spread quickly from the far north to the end of South America. The last places to be occupied by humans were the islands of the Pacific—indeed, New Zealand seems to have been reached by the Maoris only about AD 950, with the main immigration in 1300.

The full history of this epic expansion will never be known to us. The bands of hunter-gatherers left few marks on the countryside, no towns or buildings, just remains of tools and food refuse in caves and rockshelters, and some skeletons from which some of their history and society can be tentatively reconstructed. It is a pity that this is so: there can be no greater epic tale to be told of humans than the story of the first people to enter the continents of Australia, Europe, Asia, and the Americas. Their story must have been one of ultimate success. It is reasonable to guess that in the long history of the population of the world by modern humans, some bands of people must have undergone great periods of hardship and suffering; many groups must have been wiped out by cold or famine, or been reduced to a few individuals who lived and became the ancestors of new and larger bands that survived and thrived and spread out into land empty of people.

Humans started to build and live in towns and cities only a few thousand years ago, after the first of many major discoveries that irreversibly altered human existence—the discovery of how to cultivate plants and domesticate animals that happened at least 10 000 years ago in several parts of the world. Until this point, humans could never live in large groups because the pressure on the resources of the land—wild animals and plants—would have been intolerable. The discovery of agriculture altered all of this and allowed large numbers of humans to live in the same place, first in villages and later in larger settlements. We do not know how humans came together in settlements. Was it one band that simply stayed put for longer and longer periods of time or did many bands come together at particularly favourable places? In any event, the settlements must have gradually filled with humans that carried their history with them. Perhaps the telling of tales passed down the history: in most cases, we will never know. Most remarkably of all, every human also carries a history that we have learnt to read only in the past 20 years; it is a history carried by DNA.

Understanding the history of human beings seems to be a long way from what we originally were discussing—how differences within genes could be so common—but it is the history of human beings carried in DNA that is the key to explaining why differences in DNA can be so common. To understand this we need to think about the movement of modern humans out of Africa, which began maybe 100 000 years ago.

We now know that the number of people involved must have been quite small, probably in bands of hundreds rather than thousands. In times of hardship, the bands must often have been reduced to a handful of individuals and, in times of plenty, these tiny bands must have increased in numbers dramatically. To generalize, then, populating the world during the past 100 000 years must have been associated with tremendous expansions and contractions of human numbers until the discovery of, amongst other things, agriculture.

The life of humans from about 10 000 years ago became far more stable: agriculture can provide a more reliable and steady source of food than hunting and gathering, and so human numbers started to rise inexorably. Now there are about 6000 million of us. If you put this story together it means that all human beings today must share a few ancestors. How few is hard to say because we will never know the detailed history of how the bands expanded and contracted, but it is possible that in many countries all existing human beings could be the descendants of just a handful of distant ancestors. Human history is therefore the history of a few survivors of an epic journey, out of Africa, that probably started around 100 000 years ago. That expansion has left an indelible mark in our DNA; that is our history and the reason why differences in DNA are common and found in all of us.

The history of modern humans as one of expansion around the world from an African base is widely accepted today but controversy surrounds the fate of other human-like creatures that lived in Europe and western Asia between about 200 000 and 30 000 years ago—the Neanderthals. Neanderthals were short, strong, and powerfully built and they seem to have co-existed with early modern people for some 60 000 years in the Middle East and for perhaps 10 000 years in Europe until they 'disappeared' from the archaeological record around 30 000 years ago. The question that has stimulated endless argument is what happened to the Neanderthals? Did they simply vanish in the face of competition from the more successful 'modern' humans?

A fascinating skeleton of a 4-year-old child, dating from around 24 000 years ago, was found in Portugal recently; it has the hallmarks of having a mix of Neanderthal and of modern human characteristics and this has stimulated a great deal of argument amongst anthropologists. If the child is indeed a 'hybrid'—a mix of Neanderthal and modern—does this mean that the two types of human interbred? Are some of our genetic differences derived from these more ancient of peoples? Variation in DNA can contribute to these arguments and DNA from Neanderthal bones has been analysed. The best interpretation is that humans and Neanderthals did not mix: even this evidence is capable of different interpretations, however, and certainly the child's skeleton is puzzling. The arguments for and against are complex and turn on the detailed examination of the skeleton and on the method of DNA analysis used. From our point of view it is perhaps best to leave the argument unresolved because similar controversies exist in the interpretation of the archaeological record from other locations in the world.

One further tantalizing question is which humans first invented language? Whoever first did so had an enormous advantage in the competition for survival, but the sad fact is that language leaves its origins in neither the archaeological remains nor in gene differences. Clues that humans were communicating by a spoken language by at least 50 000–40 000 years ago come from the remains of increasingly complex stone tools and the first appearance of art and ritualistic behaviour around that time. It is tempting to think that the extraordinary spread out of Africa of modern humans might well have been powered by a rudimentary language ability but, sadly, there is perhaps no way we will ever know if this is true.

There are intriguing possibilities within these speculations but, for the moment, all we can conclude is that modern humans spread rapidly, possibly with the aid of language, and this expansion has left its mark in our DNA differences; it is to these, and not their detailed origins, that I now turn.

Family trees and distant ancestors

What is the history that our DNA carries? Very simply, it is the exact DNA differences in each of us. Let us think for a moment about these differences. We are given a set of DNA from each of our parents, which means that we will also be given a set of differences in our DNA: some differences will be from our mother and some from our father, who will have been given them from their mother and father, who in turn will have received them from their mother and father—and so on back into the distant past generations.

Every generation, of course, adds some differences of their own; if you recall, we all have about 30 differences that are unique to us due to failure of the exact copying process for making new DNA. But most of our differences will have been passed down the generations from parents to children and the remarkable thing is that we can look at the differences in us—the existing human generations—and work out which of these differences must have been present in our most-distant ancestors. This would be very difficult if we all had different ancestors but this is where the epic history of the human population becomes crucial.

There seems little doubt that all human beings now alive are descended from the few humans that came out of Africa around 100 000 years ago. When these humans colonized other parts of the world, they carried their genes and DNA differences with them and these differences, just as is the case for a modern father and mother and child, will have been passed to their descendants. The important question we then have to answer is what DNA differences were present in the small groups of humans that populated Asia, the Americas, Australia, and Europe? Remember, the history of these groups was one of hardship. At times the groups may have been reduced to just a few humans who survived and thrived and so we must all be descended from just these few survivors. We should therefore share the DNA differences of these individuals, even if many new differences have been introduced since those distant events.

Should we then expect the same differences to be found in everybody, the world over? Probably not; a number of small groups colonized different parts of the world but we believe that all would have a common African ancestry and so shared common DNA differences. Each band would have its own history and so it is to be expected that it would in time develop its own pattern of shared differences. What we expect to see in modern human populations, therefore, is a mixture of shared and non-shared DNA changes. This is indeed what geneticists observe but the results are not as clear-cut as you might imagine (a very important point to which I will return when I discuss human 'races'). What also emerges is that the distant ancestry of existing human populations accounts for the DNA differences that are common in present-day populations. To understand why this has to be true, we have to consider how it is that we can work backwards from the present and establish the DNA differences in our most distant ancestors.

Trees of DNA

We need to do an experiment at this point: we are going to establish the exact order of 5000 rungs from exactly the same region of DNA in 100 different people selected from all over the world—from all of the continents—and then compare the individual DNAs to identify all of the differences. To make this easy, we'll concentrate on the first 10 rungs and you can imagine this applies to all. We will also write down only the rungs attached to one rail: so rather than write the whole rungs as GC then AT then TA, CG, and so on, we will write GATC. Using this convention (which I introduced in Chapter 2), the first 10 rungs of the ladder in the first person we analyse is GATCCCGGAT; the next person has exactly the same rungs, so we write this as 10 dashes (–) to signify they have no differences, ––––––––––; the next has one difference, the first T is changed to an A (so GAACCCGGAT): we record this as A–––––––––.

We can represent all of this information as a series of diagrams where, first, we show only differences and, second, we do not include more than one example of the same rung order, so if there are actually 10 people with A–––––––––, we write it down only once. The picture looks like this:

GATCCCGGAT
––A–––––––
––A––G––––
––A––G–T––
––A––G–––A

Each one of these DNA differences is in a present-day human but the alteration from the original GATCCCGGAT must have happened at some time in the past. The question is, did any alteration happen before the others? If you look carefully at this diagram you begin to realize that there is a relationship between these orders; the alterations might have occurred in a particular order such that a DNA region with a difference was itself changed a second time.

The logic of this relationship is established as follows: the original ancestor must have had the full GATCCCGGAT, but let's suppose that in one of his or her descendants the T was altered to an A, to produce the order − −A− − − − − − −. This rung order must itself have undergone a further change to − −A− −G− − − −, which was itself altered to − −A− −G−T− − in one person and to − −A− −G− − −A in another. We know this is the case because this is the simplest way all of these changes can relate one to another—any other relationship would require the simultaneous change of two rungs, which is unlikely to happen (certainly far less likely than the consecutive order I have put forward). If two further individuals have C− − − − − − − − − and C− − −G− − − − −, then a second set of changes must have occurred: first the C− − − − − − − − −, which then changed to C− − −G− − − − − −. What we can now do is start to construct a historical family tree of all of these changes, using arrows to indicate the order of event. It would look like this:

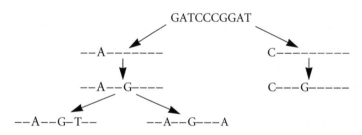

This picture is almost a family tree but many generations are missing because we cannot know how many have passed between each of the changes. Nevertheless, all of the evidence points to the overall picture being correct. So we can make some simple predictions: any person related to the C− − − − − − − − − 'branch' of the tree will have the C− − − − − − − − − or the C− − −G− − − − − derivative, whereas people related to the − −A− − − − − − − branch will have one or other of the four possible rung orders of this branch. Hence people can be sorted into different branches and groups based on their DNA differences; it's even possible to make some good estimates of how the different groups interrelate: one would guess that the − −A− − − − − − − branch has a common history different from the C− − − − − − − − − branch.

This, in brief (and much simplified), is exactly what can be done in examining existing human populations. Like almost everything to do with genes and human beings, however, the reality of doing such analyses is far from easy. In practice, geneticists do not work with just a few differences, which means it can be extraordinarily difficult to work out how the 'tree' fits together. This is not because of anything particularly mysterious but simply because the number of different 'trees' that you can put together from even a simple set of rungs is very large indeed.

The easiest way of comprehending the difficulty of the problem is to think about the base of the tree. In the example I have given above, we do not need to know that the original ancestor was GATCCCGGAT, because you can see that it is possible to work out a simplest 'tree' by looking at all of the rungs and trying to put them together into the order that requires the smallest number of rung differences for each step, working backwards until you establish the original rung order. But where we are dealing with many more differences and individuals, it is not always possible to define a unique 'tree' because there may be several that are equally coherent. Worst still, there may be so many ways of building the tree that the problem become intractable even for the very fast computers used to analyse the information, and so researchers have to rely on sophisticated mathematical analyses to produce a reasonable interpretation of the tree. Does this matter? Yes, of course, for if the base of the tree is wrong, then everything we conclude about how populations relate to each other will also be wrong; even if the base is right and some of the branches wrong this will also lead to mistaken relationships.

Drawing up family trees is not the only way to study relatedness. Another way of doing this is simply to record how often differences occur in any group of people: if one group shares a common ancestry that is different from a second group, the first group may contain DNA differences more commonly than the second, and vice versa. This allows human groupings to be defined by the differences in 'frequency' (the technical term for the commonness of a DNA difference) of differences within both DNA and proteins.

Both approaches have their relative merits and deficiencies, and consequently this field of genetics is more than usually dominated by different scientists arguing over the meaning, or otherwise, of conclusions. Sometimes the arguments are based on different interpretation of the same information and sometimes different analysis of the same population—the replication problem again. In science, disagreement is part of the process that we use to arrive at an accurate description of the reality, and so even though disagreement continues, gradually a clear truth emerges. All of the evidence points to DNA differences being shared among all human populations, with some more common than others within individual human groups, exactly as one would predict from the history of the human population. It's impossible to be completely accurate—we cannot infer the original rung order of the founders of any population but we can certainly identify some of the characteristic DNA differences.

All of this research adequately explains why different peoples from different distant parts of the world can have different rung orders, but it completely fails to explain why people from the same region can have common rung differences—such that half of a population has one order and the other half another. To understand this we have to look a bit more carefully at how human populations are made up. In particular, how do humans choose the mother or father of their children, because, as we will see, it is this that determines many of the common differences in rung order between us.

Choosing partners—home or away?

In the not-so-distant past, travel between different countries, or even between different regions of the same country, was difficult or impossible. There were many reasons for this; humans can, for example, be most unfriendly to strangers trying to pass through their lands or mountains; and rivers or lakes can bar passage. In the past, therefore, people were forced to look within their local region for a reproductive partner; this has interesting consequences for our ideas about DNA differences. The result of limiting the choice of partner to members of a small group is that after a few generations the partners will probably be related in some way: the smaller the group, the more likely this will happen.

To understand what happens to DNA differences in small groups, we have to make some assumptions. Let's start by assuming that the initial group is organized into 10 families, and each family has a different set of DNA differences (this assumption may well be wrong, of course, because the group will already have a long history, which might mean that they are already related; but we will ignore this possibility for the moment). Let's make another assumption: each family has children who are about to find partners and have children. The question is, which individuals in which families will pair off to have children? Probably it will depend on the usual events that control choice: in our hunting and gathering days perhaps strength, agility, and size were important. What is the possibility that the pairings will all go so neatly that when the children are born, the 10 original families are equally represented in the next generation? What if one family had very tall, strong children who were great hunters and gatherers? Could it be they were particularly favoured partners? Another family may have had only weak and sickly children and so no one selected any of them as partners. In both cases, the next generation will have been produced by only part of the original 10 families—the 'strong family' contributing most and the 'weak family' not at all.

What has this pairing done to the DNA differences in the group? It must have reduced them in number because if the original 10 families had 10 sets of DNA differences, then in the next generation there are no 'weak-family' differences (they had no children) but there are more 'strong-family' differences, which means that the range of DNA differences has been reduced. In each generation a similar process happens; each generation sees the loss of some individuals or even whole families—perhaps the 'strong family' were all killed when their cave collapsed on them or some other catastrophic event occurred. We do not need to know the detailed history, because the principle is always the same: in small groups, events conspire to reduce the number of DNA differences.

Reduction in the number of DNA difference in a group has two very different outcomes: a DNA difference could spread so that all of the group has the difference (we call this 'fixation') or a DNA difference could be lost from the group ('extinction'). In either event, the differences within the group will be reduced in number and, equally, the differences between different groups will be increased in the process. In practice, there will be a period of time where the DNA difference will be present in different individuals in each generation and the presence of the difference will fluctuate from generation to generation until finally it becomes either fixed or extinct.

Is there any end to this process? Surely it means that all the individuals in small groups will become identical? In principle, this could happen but in practice it's unlikely in any animal or human group because if too closely related individuals breed the effects can be disastrous (this is called inbreeding). Inbreeding comes about because two close relatives have an increased chance of having exactly the same set of gene differences. If they have children, the children have a risk of inheriting the same gene differences from BOTH parents and this can cause an actual disorder in the child; the parents, with only one copy of the gene difference, would not themselves be affected. The extreme of inbreeding is brother/sister mating and this is one reason why virtually every human society has strong restrictions against incestuous relationships.

From the point of view of the bands of early modern humans, inbreeding must have been a big problem. It was probably solved by the exchange of a small number of individuals between groups—perhaps by capture of men or women or by exchange of individuals by some other process, which would introduce a new set of DNA differences into the group and so avoid the worst effects of inbreeding. Of course, these incomers also mean that no human group is truly isolated from the rest of the human population—a critical point when I discuss the rotten foundations of the concept of human races (Chapter 5).

All of these considerations lead us to assume that there were few DNA differences in the first modern humans when they started to leave Africa and colonize other continents. As the bands spread out and set up new groups, each of these in turn would tend to contain different sets of DNA differences. With the discovery of agriculture much later, the small hunter-gatherer groups start to expand in size and this rapid increase in population changes the whole dynamic of DNA differences.

As a human population increases in size, two effects occur. First, the chance of whole sets of DNA differences becoming fixed or extinct becomes much smaller because so many people are breeding in each generation. If one weak family produces no children, the effect on a population of 1000 families (one set of differences out of 1000 is just 0.1 per cent) must be 100 times smaller than a population with just 10 (one set of differences out of 10 is 10 per cent). Second, different parts of the population might expand differentially. Perhaps there were two dominant families amongst the original 10 and these, by virtue of their position, might have bred more successfully than the other eight families; this means that the expanded population would predominantly, but not exclusively, consist of two large groups of relatives.

So in a group that has expanded relatively recently and very fast, as is the case in most human populations, what we expect to see is that all individuals will have DNA differences but the types of difference will fall into groups that reflect the ancestral structure of the population, dating way back in time to the founding hunting and gathering bands. Of course, because new DNA differences are being introduced every generation, some differences will not derive from the original founders of the population but will instead be present only in the descendants of the individual in whom the difference first occurred. If this individual lived in the distant past, then many of the current generation could be his or her descendants and the DNA difference could be relatively common in the population. Conversely, if the individual lived only two or three generations back, then the DNA difference will be present only in those few descendants in the current generation; overall, the DNA difference will be rare in the population.

We can now start to understand the DNA differences in current human populations. We see a spectrum of commonness of differences: some will be present in almost all individuals and some in a very few; those present in most individuals we suspect to be old—the relics of our ancient history—whereas those in only a few individuals are relatively new and restricted to a small group of individuals with a recent ancestor in common.

This, then, is why we have both common and rare differences in our DNA. Understanding this is important because amongst the common differences will be those that contribute to the common diseases. We are therefore faced with a quite remarkable problem: if we are to understand the underlying causes of human disease then we have to understand the history of human populations. I think that this is a delightfully unexpected connection; fortunately it is a problem that we should be able to solve.

■ Finding the common differences

If we want to understand the contribution of gene differences to any common feature of human existence, then the obvious approach would be simply to analyse all 3000 million rungs of DNA from a few thousand people selected from around the world and compare the sequences. All we would then need to identify is the rung differences common to each population and somewhere in this collection of common differences would be the differences that contribute to common features of that population. What I have actually proposed here, though, is that we run the equivalent of a human genome project on every one of a few thousand individuals. But we can be pretty sure that even a few thousand will not be enough to cover all human groups. In short, getting the information we need is going to be impossibly expensive!

Fortunately there is a simpler way of getting this information, because all we are interested in finding are the common rung differences, not the rung differences in a population. The distinction is critical: a common rung difference might be found in 1–50 per cent of individuals whereas rare differences are found uniquely in one individual or in a small group. This gives us another approach: what would happen if we mix the DNA of 10 individuals and then use a machine to determine the rung order of a particular region of DNA—say a gene we think contributes to diabetes? This is technically feasible and does not require great expense or new technology. What we would see in the output of the machine is a perfectly normal rung order—GATCGG, and so on. But what if one person in my 10 samples of DNA has a rung difference—say the T has changed to a G—so this individual is GAGCGG, whereas the other nine have GATCGG? In this case, the machine would output the information that the rung order was GA and then a mixture of T and G followed by CGG (we can write this as $GA^T/_GCGG$).

Now let's suppose that the T to G difference is common in the populations from which I have drawn my 10 samples: that 50 per cent of people have the T and 50 per cent the G rung; again the machine would give the same result. These machines, however, have an important feature: not only can they give information about the identity of the rung, they can also indicate how much is present. In the first case the machine will indicate that 10 per cent of the sample is T whereas in the second case it would indicate that 50 per cent have a T rung, so five people have GATCGG and five have GAGCGG. In this way researchers can easily detect a common difference in a mixed DNA sample and this enormously reduces the cost of the experiment. The machines are actually very sensitive and can measure with reasonable accuracy the amount of each rung in a sample of mixed DNAs down to about one difference in 20–40 mixed DNAs.

Using this approach there is every reason to hope that geneticists will discover many of the common rung differences in human beings. The experiment is being carried out as I write by a large collaborative project (with the acronym SNP) between medical research organizations, government laboratories, and several of the world's largest pharmaceutical industries, in a 100-million US dollar experiment. It is still a bit unclear as to how many common rung differences there are in our DNA—probably more than 200 000 and fewer than 3 million, but many of these are not within genes or their switch regions and do not alter proteins in any obvious way. The experiment is designed so it will identify any DNA difference, not just differences in genes. When it is finished it will have identified many but not necessarily all of the common differences in humans—surely an enormously exciting prospect and the prime reason that led me to suggest, at the start of this chapter, that this was a revolutionary discovery of science.

As always in the science of genes and humans, there is a problem hidden in all of this and that is the answer to a simple question: what population should we analyse to identify the common differences? This is an obvious question but one that has caused much difficulty because defining human populations becomes inextricably tangled with some of the most-sensitive political problems of human history—the definition of national, religious, tribal, and racial identity. It is to this difficult problem that I now turn

What population and what's an American?

The answer to the question of which populations should be studied is rather dependent on a second question—'who's asking'? This is because two groups of scientists are interested in human populations: medical researchers and historians of human origins, or anthropologists. Medical researchers are primarily interested in understanding how their own population is divided up into groups and finding the common DNA differences within groups; this is vital information to help study the genes that are different in the common medical conditions we have discussed. Anthropologists, in contrast, wish to use DNA differences to study the relationships of human populations today and their origins. At once you can see there could be considerable disagreement over the populations we should study. Medical researchers naturally will wish to concentrate on the population of their own country; this might seem self-evident but actually even this is not quite as true as you might imagine because studying some populations from, for example, remote Pacific islands may be an important way of identifying gene differences involved in diabetes and obesity in Europe or the Americas. Nevertheless, it is generally true that medical researchers in the United States or in the United Kingdom will concentrate on their own population. In contrast, anthropologists will inevitably be interested not only in their 'own' groups but in a huge geographical swathe of people; after all, the events of the past 100 000 years have covered, literally, the whole globe.

Even this picture is a bit too simple. The US might well be populated by people whose passports suggest they are 'American' but one need go back only about five generations and one would find that the ancestors of most present-day Americans are Africans, Europeans, Chinese, Asian—anything but 'American' in population terms. In fact, only about 0.9 per cent of the American population today is truly indigenous to the US; these people (called Native Americans) are the direct descendants of those bands of hunters who crossed the Bering Strait 14 000 or more years ago and colonized the empty (of humans) American continent, both north and south. Their number was enormously reduced in the violent contacts with Europeans after the first colonization of the Americas at the turn of the fifteenth and sixteenth centuries, and many more Native Americans died from smallpox and other infectious diseases caught from the invaders.

The situation in the UK is not quite as complicated: this population has a large indigenous group combined with a substantial minority of immigrant peoples who are mainly derived from the Indian subcontinent and from the Caribbean, with smaller Far Eastern groups. Even this picture is not quite accurate. The 'indigenous' British also include some quite distinct subpopulations; for example, the Celtic tribes who were particularly concentrated in Wales, Ireland, parts of Scotland, and the far South West of England—in the county of Cornwall. The Romans, Germanic Angles and Saxons, Scandinavian Vikings, and the Norman French have also invaded Britain in the past. Each of these groups brought their DNA differences with them. So even the population of a country that is relatively 'old' compared with the 'new' world of the recent American peoples is obviously complex.

With this composite of origins, the choice of which population to study at the level of DNA becomes extremely difficult. Indeed, for medical researchers it is now more difficult than for anthropologists because medical benefits must apply equally to all members of any country, quite irrespective of the ultimate origins of a person's DNA differences. In contrast, anthropologists need to target only those groups thought to be important in terms of the spread of humans across the world; for example, Europeans are said to represent a single group because they are of relatively recent origin compared with the 100 000-year-old Africans.

In the search for the common rung differences in DNA, which population will be studied in the big project I mentioned in the previous section? To start with, it has had to be a slightly arbitrary choice and in the US the researchers are focusing on a very diverse group, deliberately chosen to increase the chance of finding differences. Consequently, it has a delightfully exotic composition and is made up of 10 individuals—Amerindian, Melanese, Biaka Pygmy, American Black, Chinese, Japanese, Russian, and three Europeans. In every case, the identity of the actual individual has been carefully hidden, so that no one knows the identity of the individuals concerned.

This is not an unreasonable choice for several reasons: the US drew immigrants from almost all corners of the world and so its current population is a reasonably representative sample. Also, of course, the sample will clarify the origins of individuals who are the legitimate target of health researchers. It is not a complete solution. For example, many groups from the Indian subcontinent, from China, and from the Australasian world are virtually missing from the US population.

The information that is accumulating from these researches shows that there are common rung differences and we already know, from smaller experiments focused on specific diseases and 'candidate' genes, that common rung differences can be associated with diseases such as diabetes. All of this leads to a reasonable degree of optimism; the researchers will have identified a substantial number of common differences in our DNA within the next few years (I am writing this in 2001) and it's a good guess that the gene differences associated with many common disorders will be known within the first decade of the millennium.

Now I want to return to the idea that we all share a common history and so must all have common DNA differences in us. What this means is that people who have common ancestry are likely to have common DNA differences and perhaps will be more similar than people with other, shared, ancestry. This works quite well; Americans whose ancestors were from Africa are recognizably different in many ways from people whose ancestors were from northern Europe. Some people (I will discuss this in detail in Chapter 5) believe this means that we all fall into distinct groups of human beings who share an identity based on gene differences. What I want to show now is that this view could not be further from the truth. I want to discuss why genes make each of us unique—why in all of recorded time there will never be another you, or me.

Just how unique a person are you?

All of us would probably like to believe—with various degree of comfort and discomfort—that we are unique in this world: uniquely talented, uniquely beautiful, ugly, lucky, unlucky, stupid, clever, or a thousand different unique-nesses. Although we may be a little like many of our friends and relatives, beneath it all we are all reasonably certain we are pretty unique. I'd like to pursue this thought for a little because it relates to the fear that we are 'programmed' by our genes that I mentioned earlier. Implicit in this fear is the idea that perhaps genes force us along a path—the source of the idea that we are programmed—and that this somehow makes us predictable. By now you should be aware of why environmental influences make this fear unfounded. What I want to discuss in this section is the incredible power of genes never to produce the same result twice; this means that even if genes really did pro-gramme you, one of the really remarkable things is that gene differences would never be the same twice in any individual and so never produce the same 'programme'.

Let's think about this in a slightly more concrete way by focusing on a specific problem of uniqueness and programming. What is the possibility of someone being born into the world that looks just like you but to whom you are in no way related? This question actually contains several reasonable suppositions— that what we look like is the product of genes, gene differences, and our environment. Many of us, of course, live in generally quite similar environ-ments: we are well nourished, housed, educated—in short, the environmental effects on us are moderately similar, which could mean that if two people had the same DNA differences then they might end up looking quite similar. So the chance of two unrelated people looking identical can be considered as the chance that they contain the same DNA differences.

Unimaginably big numbers

I have to admit here that although I am not a good—or even competent— mathematician, there are times when I love numbers because they are so unexpected. One of my all-time favourite numbers is the result of working out the rough mathematical answer to the chance that two people would contain the same DNA differences. The answer is so close to 'no chance whatsoever' that I think I will call it zero, but the maths shows why.

Each of us is given a set of DNA differences from our mother and from our father and these account for the differences between child, parent, brothers, and sisters. Actually working out how many differences there are in people is difficult because, as we have seen, differences can be more or less common in different populations and we still have no easy way of counting them. Fortunately there is a slightly simpler way of working out how unique our DNA differences might be. Let's assume that our mother and father have absolutely the same DNA differences; in fact, let's pretend that every living person has the same DNA—no differences at all (this is impossible, naturally, but let's pretend for a while).

A child is, of course, made from an egg and a sperm, which must themselves have been made by cell division and necessarily involves first making a copy of DNA. What I will do here is to take the most extreme simplification of DNA differences in real people; I'm going to assume that there are only 3000 differences that are present as the result of errors in the copying process or inherited from the parents. In Chapter 2, I showed that errors in making DNA copies occur once in about 100 million rungs, and so each one of us contains some 30 errors compared with either of our parent's DNA and the rest are inherited from the parents.

So here is the core of the question; what is the chance that two babies would have the same errors in the same places? Calculating this produces some quite stunningly large numbers: the chance of an error being in the same place in two individual's DNA is one in 3000 million, but if we want 3000 differences in the same place then the number is staggering—5 followed by 19 300 zeros!

So if we are to meet somebody who has exactly the same DNA differences as ourselves, by chance alone, we might have to meet an unimaginably large number of people. There is not even the remotest possibility of ever doing this and we can show this with one last mind-numbing calculation. There are about 6000 million people alive in the world at present, so let's say this number grows to about 10 000 million—it makes the sums easier. Each generation lasts about 20 years so we can say that a new set of people, and therefore another 10 000 million new sets of DNA differences, will come into existence every 20 years. So it would take about 5 followed by 19 200 zeros of generations to run through all possible DNA differences; multiply this by 20 and you get the number of years (about 1 followed by 19 210 zeros!). I think it unlikely that Earth will exist for more than a few tens of thousands of millions of years; let's be really optimistic and say a million, million years—1 followed by 12 zeros. Now I think you are getting the message. There will never be enough recorded time for all possible combinations of DNA differences to happen, not even if the world was to survive for millions or billions or trillions of years longer, with millions or billions or trillions more people than now live in it.

The basis of these calculation is, of course, not quite accurate because many of the 3000 differences in the DNA will not affect genes and so will have absolutely no consequence to the individual that contains them. Somewhere between 1 and 10 rungs out of every 100 on the DNA ladder is in a gene—the rest is in the spaces between genes—so we can reduce the numbers by a factor of 10 or even 100, meaning we perhaps contain not 3000 differences but only 30 that would be practically important. So we can do the same calculation for 30 rung differences, which works out as a relatively larger 7 followed by 251 zeros of different ways to distribute the differences! Perhaps this means the chance of meeting your unrelated identical twin has now been increased from the never to the slightly less than never.

If you are still not quite convinced by these ridiculously large numbers, then do keep in mind that the calculations assume there are just 3000 differences, which is, of course, untrue. If we put the full 3 million parental differences into the calculation one can show it simply adds a number more zeros to the number of possibilities—in fact, another few million or so!

The detail of these numbers is not important compared with what they imply, which is that each of us is unique and if we ever feel that our gene differences are forcing our lives in some direction, we can be sure that their influence must be forcing us in a similarly unique direction. I think that this is an interesting way of thinking about the gene's effect on our existence: the gene has lost the element of predictability that is one of the core apprehensions that surrounds the uniformed view of genes as deterministic and threatening. In place of predictability and determinism, we can place the perhaps equally challenging concept of each of us as an irreplaceable individual.

In conclusion

The keynote of this chapter has been variation and uniqueness from the gene's point of view but this is just one part of the equation that affects us all. This equation simply states that gene differences + environment = the individual. Individuality is defined by the numbers of total possible gene differences and involves incomprehensibly large numbers, but even these have to be multiplied by the even larger number of environmental influences that we can experience in our separate lives. For humans, as for all living organisms, this means that every individual is unique and unrepeatable in a way that is really quite unimaginable in relation to everyday definitions of 'unique' and 'unrepeatable'.

Nevertheless, it is equally important to recognize that although many of the differences that we are born with are unique to us, we also share gene differences with our parents and they, in turn, share them with their parents—and so back into the distant history of human ancestry. Paradoxically the same processes that make us different also define kinship and ancestry. This means that the picture of human beings and genes that emerges out of our discussions is one of great subtlety: we are the product of invariant genetic programmes, upon which is superimposed the vast complexity of gene differences and environment. Paradoxically gene differences, whilst being a source of our uniqueness, are also markers of our ancestry and ancient kinship: they make us different, but they also provide some of the features that are indicators of our shared personal histories and ancestries. It is to these twin themes—uniqueness and similarity—that I now turn to understand why they are inextricably linked to our survival both as individuals and as members of human society.

Differences, genes, and lives

The future lives of Jeanne Dream and Jean Battler were moulded by gene differences and environment. Perhaps it is worth trying to consider what Jean Battler's life might have been like if she had the environment of Jeanne Dream: what would have been better, what would have been worse? We cannot know with certainty but we can guess it would have been a different and, perhaps, happier life. For all of us, gene differences are one of the determinants of the quality of our lives and this is true for every living species. More significantly gene differences are the route to death and survival in the face of a changing and often malign world: they are a key to individual and collective survival.

The history of human life on Earth is inextricably linked to individual differences, and this history is deeply entwined with the geography of human origins and of human populations and groups. Groups feature strongly in the life of Jean Battler, who, at several points in her life, suffers because it is perceived she is a member of the 'wrong' group of humans. Because of this, she is discriminated against and ultimately is killed. Grouping humans by skin colour or religion has been the source of terrible suffering in this world; do gene differences support the idea of 'races' and other kinds of human group?

'Race' has always been one of the most contentious issues within human societies and within its wars geneticists have provided some of the ammunition to fight on both sides: the battle ground of the relationship of race and intelligence has been a particularly bloody one. What is the science behind these claims? If intelligence really can be predicted by gene differences, then the social consequences that would flow from this discovery could have a profound effect, for good or for ill, on our societies. This is shown by the indelible mark the genetics of intelligence makes on Jean Battler's troubled life; it is made far more desperate because gene studies are used to show she has too low an intelligence to warrant the state wasting money on giving her a normal education. Is this really the future we face?

The area I will discuss first is how individual differences are of fundamental importance to our personal life stories. I will then move on to discuss how an understanding of gene differences can contribute to our perception of similarity within and between different groupings of human beings; such perceptions are at the heart of the idea of human 'races'. Sadly, even though modern genetics shows quite clearly that there is no scientific basis to the existence of human races, this view probably comes too late to alter the historical dimensions of the racial problem.

■ Genes, your health, and the hidden importance of being different

Human history is marked by passages of great suffering and much death and in such times survival seems as much a matter of luck as it does of design. But we are now starting to understand that just as humans look physically different from one another, so they have different abilities to withstand the challenges of disease and suffering; these differences are innate and are defined by gene differences. I will discuss how there can be a spectrum of response to change and challenge, and how an appreciation of this spectrum can dramatically alter our views of some of the happiest and most tragic events of our lives. The human body faces dramatic events during its life: resisting disease and coping with the hugely varied effects of chemicals within and without.

We will never all die of AIDS

The virus that causes AIDS is called the human immunodeficiency virus, or HIV, and it is one of the cleverest of the viruses that we humans have had the misfortune to encounter. To a virus, success is measured by its ability to invade and infect the cells of the body, to make further copies of itself, and, finally, to infect further victims. To do this the virus must first avoid the body's natural defence systems and HIV uses several skills to overcome these.

First, it infects cells that have a key role in fighting infection when it occurs; these cells are called T-cells and HIV kills them, thus incapacitating a major part of the body's defences. Second, the virus is a master of subterfuge and can hide itself in our own DNA. This unpleasant trick is possible even though the virus's own genes are made of RNA, rather than DNA. HIV makes a protein that copies its own RNA into DNA; this copy is slipped somewhere into a cell's own DNA, just like a few new genes in the cell, and the HIV genes lie dormant until at some stage in the future they switch on and make new infectious HIVs. The DNA integrated into the cell is silent and so the body's defences cannot 'see' that the cell is infected, which allows HIV to survive a long period of time. The final skill is quite unusual: the protein that makes copies of RNA has the remarkable property of making mistakes—one rung every thousand or so is actually incorrectly copied. This means that the HIV proteins are continually altering in amino acid composition and the body's defences have a very difficult time keeping pace with these changes. Even the drugs used to treat HIV can be rendered useless in the face of such rapid change: drugs against HIV work by interfering with the function of one or another HIV protein and, as the protein changes, the drugs can become ineffective.

These three skills have made HIV an extremely dangerous enemy and in the early 1980s, when AIDS was becoming widely acknowledged as a major problem, there were many dire predictions that huge numbers of humans would die of the disease. The spectre of the Black Death, the plague pandemic that ravaged Europe in the fourteenth century and killed as many as 20–50 per cent of people in different areas, was raised as a likely example. Just 20 years later, we know that these predictions were not correct for the developed world (although the disease may be far more devastating elsewhere). The question is, why not? Why did the virus not assume such epidemic proportions? There are several answers to this question. There is no doubt that the virus is somewhat less effective at infecting people, even though they have been exposed to it, than was at first feared and 'safe sex' practices have additionally slowed the spread of the virus— at least in parts of the Western world. Certainly these were important factors, but more recently another reason has been identified—the results of a magnificent piece of research, which is now a classic example of why genetic differences are important in humans.

It will help if we discuss the Black Death first because in medieval times there was no sophisticated medical knowledge that could save infected human beings: we see reality of the human condition stripped of social protection. The mortality of the epidemic was unbelievable: whole villages and communities were wiped out and yet many people survived; why did they? Why was the death rate not 100 per cent? There are some simple possibilities: perhaps the survivors were lucky and never came into contact with the bacterium, *Pasturella pestis*, that caused the disease or perhaps they caught the disease and were lucky and survived. The more complicated possibility is that perhaps some people were exposed to the bacteria but did not become infected, which might seem rather implausible. Surely a bacterium simply lives in the body and obtains its food from the surrounding cells; in terms of food for the bacterium, are not the cells of all humans the same? We do not at present know if there are human beings who are resistant to plague but in the case of HIV the answer is unequivocal; there are people who are indeed resistant to infection by the virus.

The information that led to the discovery that some people are resistant to infection by HIV emerged from a most-important task that is carried out in many countries by the medical staff who treat AIDS patients. They trace all the people the infected individuals may have had sex with and who therefore may also have become infected by the virus: these are the 'contacts' of the patient. The contacts are visited by counsellors, who can arrange tests for HIV infection, advise of possible dangers, medical approaches, and so on, and this information is passed back and kept in confidential records. Over the years, the records kept in New York have been studied by specialist medical researchers to try and understand the process of HIV infection. They reveal the type of sexual contact (whether it was homosexual or heterosexual), whether the contacts then became infected with HIV and developed AIDS, and whether they died or survived.

The studies revealed that HIV is moderately infectious—an unsurprising discovery—but also produced a startling discovery: some groups of male homosexuals included individuals who had had sex with many infected people but who themselves had not become infected with HIV. These men were tested repeatedly—in case they were just slow to develop the infection—but never showed signs of the disease. They remained something of a scientific enigma for a few years until research in 1996 showed not only that they were indeed resistant to infection but also exactly why this was so: it was because these individuals contained a small difference in just one gene.

To understand how this can happen, we have to look at a very small part of the HIV's life cycle, the moment it has to pass from the outside world into the interior of a cell. HIV cannot grow outside a cell because it lacks the machinery to make more of itself. The only way it can grow is by physically entering a cell where it can borrow the appropriate proteins from its unfortunate host. Entering a cell is not a simple operation. In the body, the preferred host for HIV is a T cell, of which there are several different types: HIV can only grow within the right sort. Fatty layers and proteins, which control the entry and exit of proteins and chemicals, surround all cells and to overcome this natural barrier HIV has a yet another clever trick up its sleeve. It makes use of a protein, called CD4, which is found only on the surface of the right type of T cell. CD4 is a signalling protein that is normally involved in the body's defence from infection: HIV cleverly attaches itself to the CD4 protein on the outside of the cell and uses it as a lever to get into the cell, thus not only overcoming the barrier to entry but also 'targeting' itself to exactly the right type of T cell.

For a long time it was thought that although CD4 was necessary for HIV to enter a cell, there needed to be another protein. For example, it was known that HIV could not infect mouse cells (which, of course, do not have human CD4) and even putting the gene for human CD4 into the mouse cell was not enough to allow HIV to enter. This implied that there must be another protein involved and, sure enough, this is indeed the case: in fact, there are at least seven of them, all signalling proteins of one kind or another. But from the point of view of resistance to HIV, the critical protein is called CCR5. HIV attaches to CCR5 and CD4 and this is what allows it to enter the cell.

Where do men with resistance to HIV come in? Very simply, every one of them has CCR5 genes that are missing 32 rungs of DNA when compared with the genes of non-resistant men. As a consequence, the T cells in the resistant men lack CCR5 protein and so the virus cannot enter the cells—hence the resistance to HIV infection. Surprisingly lack of CCR5 seems to have no other effects; the protein has a function that seemingly can be lost without harming the rest of the body.

Once this work had identified loss of CCR5 as the cause of HIV resistance, researchers then looked at the genes of large numbers of men and women. These studies showed that the loss is not restricted to just a tiny number of individuals and is actually quite common. For example, in Finland about 16 per cent of the CCR5 genes have lost the 32 rungs; in Sardinia, the figure is 4 per cent; and in the Middle East and the Indian subcontinent, the proportion varies from 2 to 5 per cent. To be resistant to HIV, both of the copies of the CCR5 gene must be defective and this is found in at most 2 per cent of the Finnish population and more rarely elsewhere.

Does this mean that people with defective CCR5 genes will never catch HIV? Certainly they are much less likely to, but you will recall that HIV is itself changing very fast; we now know that some HIV strains use one or other of the six other proteins known to associate with CD4 and that allow the virus to enter. It is thought that a tiny number of CCR5-based 'resistant' individuals become infected with these new strains of HIV. Overall, though, we can be fairly certain that a proportion of the human population will remain naturally different from the rest where HIV infection is concerned. This does not mean that HIV is not a major and continuing problem—we are talking about just 1 per cent or so of the world's population being relatively safe from the ravages of the virus and this is not a comforting thought for the 99 per cent who are at risk.

Live mice and dead humans—the case of the resistant mice

The natural resistance of some humans to disease is not an easy area to study, for the simple reason that it is ethically wrong deliberately to infect volunteers with a disease if this is likely to damage them irreversibly. Almost by definition, diseases that cause the greatest harm are those that researchers most need to work on, so much effort has gone into finding animals that might be used as 'models' of human diseases. In the circumstances, infecting an animal with a disease is the lesser of two evils when the alternative is to contemplate deliberate infection of a human being. One of the commonest experimental animals is the mouse. Mice are a central part of medical research because, as I shall explain in Chapter 7, a strain can be inbred until all the individual mice are essentially identical—they have an identical set of gene differences. For many years geneticists have worked on comparing different strains of mice precisely because each strain carries a different set of gene differences: dissimilarity between two mouse strains is potentially due to the gene differences that are found in one strain and not another, and this opens up the experimental possibility of identifying the genes.

Mice, just like all animals, get infections, some of which are fatal, and so breeders go to great lengths to make sure their mice are as resistant to as many mouse diseases as possible. One way of doing this is to test different strains of mice to see if they are all equally sensitive to the common disease-causing organisms. These experiments show quite clearly that different strains differ dramatically in disease resistance, which is important information for mouse breeds but potentially even more so for human medical research.

One gene, tuberculosis, and sleeping sickness

At first sight, resistance to a disease in a mouse seems an unpromising place to start to understand diseases in humans. But a more careful understanding of the disease-causing bacteria involved shows why this view is incorrect, because mice can be infected with organisms that are closely related to those that infect humans. This is best illustrated with the example of the organism called *Mycobacterium* that causes the human disease of tuberculosis.

Tuberculosis, or TB, killed over 1.8 million people in 1998 alone—most of them mainly in the poorer parts of the world. TB was also a killer in Europe and the Americas as recently as the first half of the twentieth century but by the 1960s vaccination and treatment had almost eliminated it as a common disease. Unfortunately this was a short-lived medical triumph because new strains of the TB-causing bacterium have emerged in the past decades that are resistant to almost all of the chemical agents used to destroy them. Many people also became complacent and stopped getting their children vaccinated; after all, what is the point of vaccination if the disease is rare? This allowed TB to stage a dramatic comeback and it is now starting to regain the ground it lost; it is flourishing particularly in disadvantaged societies.

So the discovery that different strains of mice are sensitive or resistant to types of mycobacteria was important because it presents us with a unique way of understanding which genes are required for the bacteria to infect an organism. If we can understand this process in the mouse, then we can think of new ways of blocking the mycobacteria's attack that might be applicable to human beings. An obvious challenge is therefore to identify the gene differences responsible for resistance or sensitivity in the different mouse strains.

If we compare two mouse strains, we will find they have many different proteins, and so the obvious first question is to decide if the disease resistance is the result of just one gene difference or of the combined effect of many. This is easily done. All you do is breed the resistant and sensitive mice together and test their offspring: if there is just one gene difference involved, then there will be a simple pattern of resistance or sensitivity in the offspring. In contrast, if there are a few or even many genes, the pattern of inheritance will be much more complicated, with resistant, partially resistant, and sensitive mice being produced. When mice strains are bred in this way, the results are quite clear—there is just one gene involved.

Attempts to identify the gene difference were finally successful: the gene encoded a protein called Nramp and the resistant and sensitive strains of mice differed by just one amino acid. The Nramp protein appeared to span the membrane of cells called macrophages, which fight invading organisms by engulfing and destroying them. The challenge now is to understand the role of Nramp in this process and to understand how the one gene difference can have such a startling effect.

Once the Nramp gene was identified, several other pieces in the jigsaw puzzle of disease resistance in mice fell into place. It turns out that the difference within Nramp also makes mice resistant to infection by a single-celled organism called *Leishmania*, which in humans causes a very unpleasant disease called leishmaniosis. With some 12 million people worldwide infected with this disease each year, the research on the function of Nramp assumes even greater importance. How this knowledge can be transferred to fighting human diseases remains problematic: at present, we have no easy answer. Differences within Nramp in human beings do not appear to be common and so we have to try and understand its role in macrophages. If we can establish this, it is likely we will have identified a pathway of functions that we know, because the Nramp change in the mice influences disease sensitivity. Perhaps then we can interfere with the normal operation of the pathway by using drugs and so make everybody resistant to these organisms, but it is a long and tortuously indirect path to drug development.

The case of the Nramp protein is slightly unusual because it is rare for a difference within just one gene to have a profound effect on any complex process, and developing a disease is a very complicated process indeed. Entry of the disease agent into the human body—the traditional moment when we naively think we have 'caught' a disease—is just the very first step; it has to be followed by successful colonization and reproduction of the disease-causing organism, which in turn alters the behaviour of cells; only then does it have an effect on the body and disease symptoms develop. All of these processes must, inevitably, involve many genes, many proteins, and many cells, and as a consequence disease resistance is much more likely to associated with differences within many genes, not just the one. That this is true is shown by another case of a mouse, resistant to the mouse version of the human disease of malaria.

The mouse that doesn't need gin and tonic

We met malaria in Chapter 3. The standard treatment for the disease is to use drugs to kill the *Plasmodium* organism while it circulates in your blood, but these are not always effective—the organism has become resistant to many of the commonly used chemicals. Historically the first of the drugs used to treat malaria was a compound called quinine, which today is more commonly found as a component of tonic water—taken by many of us, myself included, with a penchant for gin and tonic. It has been a standing joke for many years that the only reason anyone ever drinks 'G and T' is simply for 'medicinal purposes'—to keep the malaria fever at bay. Malaria is not a joke, though; the disease kills between 1 and 3 million people each year, mostly young children, making it the leading cause of death from an infectious disease.

This sombre statistic puts a perspective on the importance of discovering a cure or a preventative treatment for malaria. Research on this problem received a huge boost when it was discovered that particular mouse strains are naturally resistant to infection by a close relative of the malaria-causing *Plasmodium*. Breeding of mice that are resistant or sensitive to infection, just as I mentioned for TB resistance, shows that resistance to malaria is more complicated than the case of the single Nramp gene. There seem to be a number of gene differences in play and this is where the different mouse strains become an invaluable tool. The resistant mouse strains were not the product of a deliberate programme to create them, and so malaria resistance must be a product of differences within genes that exist for some other reason. This raises an interesting possibility: if resistance is the product of chance gene differences, then do all resistant strains have the same set of gene differences conferring resistance? The answer that has emerged is that different mouse strains have overlapping sets of gene differences that account for the final resistance.

This was shown by two separate groups of scientists, one in Australia and the second in France, who in 1997 carried out experiments along the lines of those I've described above for TB but with the critical difference that while both used a strain called Black 6 as the resistant mouse, one group used a sensitive strain called A/J and the other two sensitive strains called C3H and SJL. The experiments involved breeding and linkage analysis, much as was done for TB, to show where the resistance genes were located in the mouse DNA. All three mouse strains had a change in a gene located on fragment 5 (mouse cells have 19 DNA fragments), but strains C3H and SJL had a gene difference within fragment 8 and only the A/J strain had a gene difference on fragment 9. These experiments suggest that at least three gene differences are involved in *Plasmodium* resistance and that different strains carry different changes.

The evidence also suggests that these three gene differences are only the most visible of the gene differences that cause resistance; others must be involved that contribute smaller amounts to the overall picture of resistance. Up to the time of writing, these experiments have not revealed exactly which genes are different—the step from linkage to finding the gene is difficult—but we can assume it won't be long before this happens. Of course, just as for TB, there is still a long way to go before knowledge of the genes' identities will lead to new preventative measures and treatments in humans.

Know thy enemy, as thy self?

At present, there are relatively few cases where we understand the actual basis of human resistance to a disease: HIV resistance is probably the best understood.

There are also examples of the reverse of resistance—extreme sensitivity. The classic case of this is caused by the loss of the body's main system for fighting infections and invasion. Our bodies are under continual challenge from the outside and we have a sophisticated defence mechanism to fight off countless attacks on a daily basis. This is an extraordinarily difficult task for the body to achieve, but just how difficult may not be immediately obvious because the problem is not just how do we fight an enemy, it is actually much more subtle: how does my body know who is the enemy?

Why is this a problem? In part because the body cannot be told by some central government that today's enemy is the people of the neighbouring country or the members of some group. The defenders of the body, somehow, have to 'know' that they have encountered an enemy. How is this done? One way at first sight seems quite fantastic; this is for the defenders to have a complete list of all the shapes of the body's molecules so that when they encounter a shape that is not on the list, they will know they have encountered an invader and can then kill it. This seems an absurd idea: how could cells have such a list? Absurd or not, this is exactly how we defend ourselves. There is no readable 'list', of course, but instead there is a complex system of molecules and cells whose main function is to recognize 'self' and 'non-self'. Our defences really do have the capacity to recognize all of the molecular shapes of our body—a startling achievement given the complexity of a single cell, let alone the million million we contain. How this extraordinary feat is achieved and how it is not necessarily infallible is what I discuss next.

In the earliest years of a newborn human life, a particularly critical series of events occur because, from birth, babies have no natural defence systems and are reliant almost entirely on protection received from proteins in their mother's milk. Even though the babies have no active defence, the baby's defence cells are nevertheless busy developing the final list of 'self' shapes. How they do this will become a central part of understanding why some babies are born with an unbelievable sensitivity to infection.

The key component of our natural defences is a series of proteins, called MHC (major histocompatibility complex) proteins, that are present on the surface of particular cells of the defence system. Their function is to break up proteins or other chemicals into smaller fragments and, literally, 'present' these on their surfaces. The role of the MHC proteins is critical: the fragments stick to the MHC proteins on the cell surface and then other types of defence cells, called 'helper' cells, use special signalling proteins on their surface to recognize the combined MHC and fragment. The resulting wave of changes within the helper cell programmes it so that it will 'remember' the shape it has recognized; if this happens in a baby, the fragment becomes part of the list of 'self' shapes. Imagine this process occurring in millions upon millions of cells with millions of fragments, and you can see how the body in time can catalogue all of itself as being 'self'. Incidentally it is the MHC proteins that are responsible for the rejection of transplanted organs by surgery and getting an exact—or near-exact—match of MHC gene differences between the human donor and the acceptor, plus powerful drugs to suppress the effects of any slight mismatch, is the key to successful transplantation.

In adults, the same process of presentation of fragments and MHC on the cell surface occurs and these are also recognized by helper cells, but now a new process can happen. The helper cell can compare the shape it 'sees' with the memory of all self shapes; if it does not recognize a fragment as 'self', this becomes the trigger for cells to attack the fragment wherever it is found. If the fragment is on the surface of an invading bacterium, then that bacterium will be destroyed by cells homing in and 'eating' it.

This is not the only defence system in our bodies: specialist cells also produce special proteins, called antibodies, that can recognize a fragment on some invading organism and stick to it, and by doing so target it (and whatever is attached to it) for destruction. Antibodies are produced by the body for years after an infection has passed and this means that the next time the disease organism invades us, it is immediately met with a defence, without having to go through the self/non-self check. As soon as the invader enters our body it is killed by antibodies lying ready and waiting in our blood. This state is called being 'immune', because possessing such antibodies renders one immune to the attack.

Perhaps the most difficult thing to grasp from what I have described so far is how it is that the helper cells can recognize the shapes of the signalling MHC and fragment so accurately. There must be literally millions of combinations of MHC and fragment. In Chapter 2, I took care to point out that there has to be an exact match of the shape of a signalling protein to its partner and also that the shape of a protein is uniquely defined by its amino acid sequence. This leaves us with a very major problem: there must be millions of 3-D shapes of MHC and fragments, so we must also need millions of different helper-cell signalling proteins—one for each combination of fragment and MHC. Of course we only have 50 000 genes, so where can we find the millions of genes needed to code for the millions of signalling proteins on the helper cells?

The way the body has solved this apparently intractable problem is quite exquisite in its simplicity. Our cells contain only a few helper-cell signalling-protein genes and each one encodes a different, but related, signalling protein. To generate the enormous diversity required, the helper cells simply break up some of the genes and join them back together in different orders so that each rejoined combination has a slightly different mixture of amino acid sequences and hence a different 3-D structure. For this reason, helper cells, and some others in the defence system, are the only cells in the body that do not have exactly the same genes as the rest of our cells. Even this mixing of genes is not sufficient to give the diversity of amino acid sequence and so the helper cells have an additional ability and can deliberately change a few DNA rungs within the genes to make signalling proteins with wholly novel amino acid sequence.

Bubble babies

The two processes of gene rearrangement and generating change means that helper cells can make millions upon millions of different signalling proteins, simply by breaking the genes, joining them together in different ways, and sprinkling in a few changes; this is the heart of the body's defence. A key protein component of the machinery required to accomplish the complicated cutting and joining of signalling-protein genes is called ADA, and we know this is key because of some very unlucky babies, mercifully rare, who are born with a defective ADA gene that cannot make ADA protein. The loss of the ADA protein means that the babies completely lack a defence system because all of the DNA joining and rejoining is missing. They survive for a time—just as all of us survive our earliest days without our own defence system in operation—but in the end the slightest infection, one that would pass unnoticed by you or me, will overwhelm them.

ADA-deficient babies have to be protected against almost everything in the outside world because almost everything carries bacteria or other living organism that would grow on or inside our bodies were it not for our defence systems. The only way of preventing massive infection is by keeping the babies inside a plastic bubble that can be kept completely free of all sources of infection.

These bubble babies show us that we are entirely dependent on our defence systems because our bodies are under continual challenge from the outside world, and, in turn, our defence system is critically dependent on its ability to define self and non-self. The need to define self/non-self means we could be open to attack via a very clever route: what if an enemy could develop an exterior that looked like something we recognized as self? Our defence cells are programmed not to attack self and so the invading organism would be free to roam through our defenceless body: this would seem to be a very considerable flaw in the system. In principle, such a disease organism could devastate the world's population because all humans have much the same proteins and so much the same 'self'. Remarkably, such organisms do occur in nature but the reason they do not destroy us provides one of the best examples of why differences within genes are essential to the continuing survival of the human species. To understand this, we need to look carefully at the MHC proteins.

The enemy within—a gap in our defences?

At the heart of the defence system is the fragment of a foreign organism presented on the surface by the MHC proteins that can subsequently be recognized as a non-self 3-D shape. One small detail in this process is of defining importance: 'self' to a helper cell is the 3-D shape of a fragment combined with the MHC protein and not just the shape of the fragment alone. This apparently minor detail is why we need not fear that an organism will mimic 'self' and devastate the human species. Let's consider a virus that on its surface has proteins with the same 3-D shape as a fragment of our own cells combined with an MHC protein, and so looks exactly like 'self'. Given that all human beings are made of the same set of proteins, it is likely that we will all have the same set of 'self' shapes. As I have said, such a virus could spread from person to person, unhindered, because all of our defence systems will be fooled into thinking it is 'self'.

This is an alarming prospect but it will not happen. Why can we be so confident? Because the MHC genes are rather unusually variable: the same genes in different individuals have numerous rung differences and this means that in most of us the MHC proteins will also have a slightly different 3-D structure. From this it follows that even though the fragments from all of us are identical, the MHC protein's 3-D structure will be subtly different and so each of us recognizes a slightly different 3-D shape of fragment plus MHC protein, even if the fragment is identical in every case. No matter how well an infectious agent might mimic the 3-D shape of a fragment plus a particular MHC shape, this 3-D structure is only recognized as 'self' by those few individuals who share exactly the same MHC gene differences. From this it must follow that no organism can ever be recognized as 'self' by all human beings.

The differences within our MHC genes are absolutely critical to our continued survival as a species. That this is so is proven by studies of tribal populations that have thrived isolated from the outside world for many generations and that have only recently been contacted by people from outside of their tribe. Papua New Guinea contains the most remarkable example of this: there are many hundreds of tribes or clans in this rugged country, who, even today, speak some 750 different languages. The geographical barriers of ravines and mountain ridges, combined with mutual incomprehension, results in a great deal of forced intermarriage within each group and, therefore, the changes in the MHC proteins are much less diverse than one might find within occupants of the larger world outside. In some cases, just three sets of MHC gene differences are present and this does mean they are open to an invader who can mimic some or all of the 'selfs' within the tribal group.

The New Guinean groups are, of course, safe while they are out of contact with the rest of the world, simply because they will not be exposed to most of the challenges that we people who live in the crowded West experience every day. The devastating effects on hitherto isolated people of first contact with the outside world have been well documented in New Guinea in recent times and, historically, among the American Indians, whose populations suffered far greater losses from diseases brought with the European colonizers than from direct attack. Indigenous people are in double jeopardy because not only are they exposed to new disease-causing organisms never encountered before by their defence systems and against which they have no antibodies, but they are also more vulnerable as a group to the invader who looks like 'self' because they have so few MHC types.

The variation of the MHC proteins is important to survival of human groups but, of course, some disease-causing organisms will be able successfully to mimic 'self' in small numbers of people of the right MHC type, who consequently will be vulnerable to attack, whereas others are resistant. The cost of fighting disease is high to humans: variation in effectiveness of response because of the MHC variation means that there will always be individuals who will mount better or worse defences against infection. Therefore, there will always be those who survive as there will always be those who die when faced with identical challenges—it is an intrinsic feature of our bodies' response to the huge pressures from invading organisms. There can be no better illustration of the need for differences between human beings, and the consequences of such differences: the death of one person is a necessary price to pay for the survival of another.

This difficult thought is at the heart of much unhappiness and misunderstanding over the idea that there can be medicines that are truly safe for all human beings; it is to this area that I now turn.

Dead from 'safe' drugs?

I hope that I have convinced you that differences between human beings are widespread and of great importance. The consequence of these differences can literally be life or death, and until we have a much greater understanding of the extent of differences we will remain vulnerable to their sometimes tragic consequences. That this is true is sadly illustrated by our response to unexpected deaths of people—often in the prime of life—after they have taken drugs of one sort or another. The emotion of these events is, of course, overwhelming: probably the most common feeling is that there must be a reason for this tragic death and someone must be to blame.

Sometimes there is indeed someone to blame: illicit drug dealers do sometimes sell lethal combinations of drugs and impurities that can kill, and even large drug companies have on rare occasions covered up the dangers of a drug they are selling (the case of thalidomide comes to mind). But the sad truth is that sometimes nobody is to blame. Deaths can occur simply because the unfortunate victim is particularly sensitive to the effects of the drug and so, even though they have taken exactly the same dosage as everybody else, their body responds catastrophically. There is now good evidence that people—much as in every other feature of their lives—differ in their response to many drugs and that these differences are in part controlled by differences within genes.

It might seem strange that the body can vary in its response to substances that have been invented by humans: how could the body 'know' what is going to be invented by the drug companies so have the proteins ready to modify the drug's actions? Why does it have proteins that can perform these functions? The answer, of course, is that the proteins that influence drugs are there for different reasons and the influence on drugs is simply a side-effect. Quite naturally the body contains proteins that search out and change chemicals in our food, and we need these so that we can digest food. Plants, in particular, contain many chemicals that are poisonous and protect them from being eaten by animals and insects; these chemicals are the plants' final defence. Animals need to eat plants and so an arms race has developed between plants and animals: plants make poisonous chemicals and animal cells contain 'modifying' proteins whose role is to destroy the plants' poisons. Without these proteins, much of the plant food we eat would be dangerous.

Plants make many different poisons and so our cells contain many different modifying proteins. An animal, of course, cannot 'know' what poisonous chemicals a plant will make and so there is no question of each modifying protein having only a single chemical target. Instead, most of these proteins recognize just a small part of the chemical structure—for example, a cluster of a few carbon atoms—and whenever the protein finds a chemical with this structure within it, it will attack, by breaking one of the chemical bonds in the cluster, or other means. This will break a poison into two fragments and so render it ineffective; it is no longer the same single poisonous chemical but is instead two smaller, and harmless, chemicals.

This means that any chemical with a cluster of carbon atoms that is recognized by a modifying protein will be broken down and this applies equally to poisons, drugs, or any other chemical that finds its way into our bodies. So when we swallow any type of drug we have to establish a delicate balance: we have to take enough so that even though our modifying proteins are breaking it down there is nevertheless sufficient unmodified drug to have an effect on its target. If we take too little, the drug will be ineffective; but if we take too much, then we might cause an overdose and suffer the ill-effects of this instead.

Perhaps the best illustration of this balance is the case of the anti-high blood-pressure drug called debrisoquin that works by interfering with the way the body controls its blood pressure. Debrisoquin is a small chemical with a group of carbon atoms that is recognized by a modifying protein with the cumbersome name of cytochrome P-450IID6, which I will abbreviate to P-450. P-450 protein recognizes the chemical shape of debrisoquin, breaks its chemical structure, and so renders it ineffective, which is critical for the drug's medical use; if it was not broken it would simply carry on lowering the patient's blood pressure until it was so low that coma and death would follow. We know that this can happen because it turns out that about 5 per cent of us have a single rung difference in the P-450 gene, which results in the protein being far less active against debrisoquin. The effects are extremely alarming; the blood pressure of these people drops to such catastrophically low levels that they become unconscious and would die without treatment.

What are we meant to make of this? Is debrisoquin a bad drug? No, in 95 per cent of cases it is a very good drug, it is simply that 5 per cent react differently to it. Now we know about this change, great efforts are, of course, made by all drug companies to ensure that their drugs are not sensitive to this common variation in P-450. This is not the only protein involved in breaking down drugs. Rare differences almost certainly occur in one or other of the many different modifying proteins in our cells. The drug companies cannot easily predict how these might affect people who are taking their drugs. There are two good reasons for this. First, we simply do not know about the rare changes precisely because they are rare and, second, why should a drug that can cure 20 000 people be prevented from use because it could harm one person?

The truth is that any drug may harm a few people. The question is how many people does it have to harm or kill compared with how many it helps before we see this as unacceptable. Everybody who has faced the sudden death of a friend or relative under such circumstances seeks to explain, to understand why death has happened; it is a natural human emotion. In the case of death from taking a drug, there is a ready explanation to hand: the drug company must be to blame for selling a dangerous drug. The truth is more complex than that, however, as there can never be a completely safe drug and there probably never will be—because of the variation in our genes. All we can hope for is that scientists will in time develop ways of testing and allowing for some of the most important variation.

To summarize, then: some proportion of the human population is very sensitive to particular chemicals. As further examples, we know that many people are very sensitive to commonly used antibacterial chemicals such as penicillin; some people are acutely sensitive to poisoning by nuts in food; others respond badly to taking aspirin (about 1 in 20 000 people); and some are extremely sensitive to anaesthetics and can die if given injections in the course, for example, of simple dental treatment. We accept these types of sensitivities and take precautions against inadvertent exposure to the substances that cause the problems, but this is only possible when we know of the danger. The tragic truth is that if 100 000 people take a new drug, be it a new medical drug or the latest recreational drug, then a few of them will respond badly because of their gene differences: some may even die but we will not know in advance who is at risk, and who is not, until the tragic 'experiment' has been inadvertently carried out: by then it is too late.

Geneticists are studying this problem with great intensity: the hope is that understanding the relatively few bad responses will enable us to identify what it is that makes people so sensitive. But the difficulty of such studies should be obvious—humans are not experimental animals that can be dosed up with chemicals to find out what happens. Consequently the causes of most sensitivities, even those to aspirin, which is relatively common, are still not understood.

Death from reaction to medical drugs is remarkably common—over 50 000 people die each year in the USA. But it is unlikely that there will ever be a 'safe' drug, if by safe we mean that it will never cause harm to anybody who takes it. A new role for the drug companies is, increasingly, to identify the gene differences that define particularly sensitive people and develop tests that will prevent them from being given potentially dangerous drugs. This research will take years, not months, and in the interim it is important that we recognize the difficulties of drug development. Perhaps the most important immediately beneficial development would be to eliminate the element of blame and negligence that is implicit in the inevitable court cases that arise from drug deaths. I suspect this is one area where government intervention into compensation would be enormously valuable, particularly if this was combined with education as to the genetic consequences of taking powerful chemicals into our bodies.

There will always be survivors, there will always be victims

The lesson that emerges from what I've been saying so far in this section is quite simple: there will always be survivors and because of this there will always be victims. As a consequence of this rather brutal truth, a great deal of effort is going into understanding who in our society is at risk, and from what, and who is not, because if medical researchers can identify those most at risk, they can prevent disease, rather than be forced to treat it. The saving in human misery will be enormous and, for once, this will also be the cost-effective solution.

We remain surrounded by gene differences and we are surrounded by challenges to our bodies from viruses, bacteria, and chemicals; common gene differences are a direct reflection of this fact. Without differences in our genes, the Black Death might have reduced the population of Europe to levels that no longer could have supported collective life; under these circumstances, the history of the past six centuries could have been totally different. Perhaps it would have been one of nomadic tribes and bands moving across the empty face of Europe, or perhaps it would have been one of the colonists from the Americas, who sailed across the sea to the sparsely inhabited European world.

Without gene differences, the human species might never have emerged and certainly would have had a different history. As I have shown, human history has also defined many of the gene differences contained in our different populations. It is to this that I turn, because the distribution of gene differences across different human populations is clearly a mark of the populations' origins and separate histories. Can we therefore use gene differences to inform our understanding of the differences between groups of humans? This is one of the most politically sensitive areas of gene science because it has bought us to the edge of an abyss—the subject of human groups and 'races' and their horrifying history.

■ 'Races': genes as excuses

The history of the study of genes is far from honourable and perhaps the question of 'race' is one of its least-honourable areas. This is a pity because genes have much to say about 'race'—primarily that human 'races' do not exist—even though, in the past, our understanding of genes has often been used in quite the contrary direction and been exploited to excuse some terrible abuses.

Race is a topic that is extraordinarily difficult for a geneticist to discuss for several good and bad reasons. A bad reason is that the self-censorship of liberals sometimes leads to the suggestion that 'race' is simply not an acceptable topic for scientific thought. As you will see, I firmly believe that this is a mistake because the science of genes has a great deal to offer on the subject. A rather better reason is that nobody seems able to define 'race' and the term will have a very different meaning to a white academic in a university in London or Australia than it will have to an African immigrant living in a cold northern European city or to a person living in the deep south of the USA. If one cannot agree what 'race' means, then discussions are difficult and often unproductive, because neither side can quite decide what it is that is being argued about.

In any event, the history and the current views of 'race' cannot easily be dissected out from that of racism, and all of the suffering and hurt that it has caused, which means that all discussions must take place against a background of history, politics, and social policy. We have to go back only 60 or 70 years to enter a time when the definition, or otherwise, of 'race' was a key to the barbarities of Hitler and the Third Reich. I may now be able to see the definition of 'race' as something I can discuss in this book, but my views, 60 years ago, would have to be in a completely different frame, a frame that, quite literally, would have been a matter of life or death.

In short, 'race' is a difficult arena into which geneticists stumble. That being said, I propose to do just that because I think there is a truth that emerges from a scientific view; I do not in any way flatter myself, or other geneticists, into thinking this truth is going to solve any of the great social and political problems of 'race'. Fundamentally the complex history of 'race' has assumed an importance that is probably greater than any scientific opinion of the reality of human differences, but we can at least strive to make the underlying facts, such as they are, accessible and intelligible to as large an audience as possible.

How can we discuss 'race'? Primarily, by trying to think about the problem from the gene's perspective. But before we do this we need at least to produce a working definition of what we mean by 'race'. I do not pretend that what I am going to use will be accepted by everybody, but I believe it will cover enough of the field to allow me to discuss 'race' coherently.

My working definition of 'race' is as follows: 'a human race is a subdivision of the human population that is characterized by specialization to different environments'. This sounds complicated but is a definition based on a simple, indeed over-simple, view of human biology that can be stated in the following way. Human populations have separately existed in geographically isolated parts of the world for many generations, and as a result have been exposed to different environments: these environments have placed pressure on the humans that live in them and, as a result, each population has gene differences that enable it to cope better with the conditions it faces. So by this argument, the white 'races' of northern Europe are white because of the sunlight and vitamin D argument, and the many other possible reasons for adaptation to the cold and harsh northern environment; the African 'races' are black because of the need for protection against the intense light of the tropics; and so on. What this definition allows for is human 'races' to emerge over a period of time, the properties of each 'race' being characteristic of its particular geography and history—or even its breeding preference because central to this idea is the view that 'races' do not breed with other 'races'. If they did, racial differences would disappear; if black and white interbred we would all be brown. This will turn out to be the key concept of our discussions about 'race'—the central importance of racial 'purity', which is completely tied up with the allied concepts of racial identity and exclusivity.

Who can join the club and what are the rules?

These racial ideas are based on an element of reality. We have seen that there are indeed human populations that have been separated over quite long periods of time and these populations have become quite recognizably different. If so, then why have I suggested that 'races' do not exist? The answer lies in understanding exactly what is meant by membership of a 'race'; what do you have to have, or not have, to be a member of a 'race'? If we look carefully at this simple question we can understand why 'race' has no basis in science.

The problem is that membership rules for belonging to a 'race' are a little vague. I'll try and illustrate this by using my own background as an example. In my imagination I am going to try and apply to become a member of the English 'race'. The rules of this 'race' would appear to be pretty simple: you must be English and your parents must similarly be English, and they, in turn, must be the son and daughter of yet more stout English folk, and so on—all of this to ensure you have inherited a good crop of solid English gene differences. Of course, I have already made one error here: historically 'race' theories rarely talk about genes and rather prefer the words 'blood' and 'bloodlines'; I have used gene where 'blood' would have been more common. In racial jargon, my 'bloodline' can be thought of as the sequence of my English ancestors and my 'blood' is what I have inherited from them through my parents, but I choose instead to talk about the reality of my gene differences.

So, my membership application in hand, I stand in front of the membership acceptance committee of the English race club to be questioned as to my eligibility to join. Question 1: place of birth—answer, Milan, Italy. Surely I am applying for the wrong club; perhaps it should be the Italian race? No, place of birth doesn't matter, provided my 'blood' is English. Question 2: origin of parents—answer, both born in London, England. The committee is pleased at this and moves rapidly to question 3: origin of all four grandparents—answer, Dorset (the West of England), Wales, England, and Scotland. The committee breaks out in total uproar: I am no English man, I am some sort of mongrel mixture of Welsh, Scottish, and—sad though they think it is—polluted English blood. My application for the English race is roundly refused.

This is, of course, a tremendous setback to my desire for social acceptance, which requires I be a member of some group or other. But I quickly recover my spirits and go next door to the British race club and they take me in with open arms, for Welsh, Scottish, English, and Irish are all quite acceptable members of the British race club.

Let's look at all of this from a few different points of view. Why was my birth in Italy not grounds for refusal? Because, of course, my 'blood' was not Italian, my 'blood' being defined by my parental 'blood' and theirs by their parents, and so on back in generations, just as my genes are. What if all my ancestors were born in Italy? Then, of course, they must be Italian and so my 'blood' is Italian. But hold on, how many times must my ancestors be born in Italy before I am Italian? This is a very difficult question: the British race club will definitely have to pass this one to its racial purity subcommittee because it is not clear exactly what 'blood' means in this context. So, part of 'race' purity must include geography. Why? Because the original idea of 'race' is itself firmly based on geography and the effects of different environments on human evolution.

Exactly what is it about my Welsh and Scottish ancestry that bars me from being pure English? The answer is, of course, 'blood' purity—I am an unacceptable mixture. Here we have just hit a real problem with 'race' theory; exactly what is pure 'blood' in these terms? Blood is a measure of origin so we can simplify all of this and say that 'blood' ought to be roughly the equivalent of the collection of genes that we have inherited from our ancestors. My genes will contain changes that ought to be both characteristic of, and exclusive to, my 'race', and because this should be a product of a particular environment we can conclude that the gene differences specific to my 'race' must also be specific to a geographical region. If the 'race' theory is correct, English people must have gene differences that make them uniquely English because they are all descend- ants of the original band of humans that first arrived in the misty and cold islands in the North Sea; for once, this is a clear description of the relationship of genes to concepts of 'race', which can be tested by experimental observation. If human 'races' exist, we can prove this by identifying gene differences that are found only in a single, geographically defined and limited human group. So, let us look for these gene differences.

Purity and exclusivity

Do such gene differences exist? There have been many studies where geneticists have analysed DNA differences in different populations and all of these studies yield the same message: no group of DNA differences is present exclusively in one human population and not in any other. Here is the reason why: the time we have been living in separate populations is short in terms of human evolution, in most cases much less than the 100 000 years or so that have passed since modern humans spread out of Africa, and so relatively few DNA differ- ences have occurred in any single population. People of ancient African ancestry contain more DNA differences than members of any of the 'younger' popula- tions found elsewhere in the world, but the DNA of even these 'oldest' human groups are less variable than that of almost any other species of animal. The relative youth of our species means that there has been insufficient time for large numbers of DNA differences to have separately appeared within different populations, and this, in turn, means that all populations share some of the DNA differences that must have been in our most-distant ancestors. DNA differences are often more common in one population than another—as, for example, must be the gene differences that cause skin coloration to be dark or light—but we cannot identify DNA differences that are present in all individ- uals in one population but are missing in all individuals of any other popu- lation.

   What this tells us is that the main requirements of racial membership— 'purity' and 'exclusivity'—are going to be extremely difficult to fulfil because there seem to be no absolute differences between human populations. The message of the genes is clear: racial groupings do not exist in humans based on the exclusive possession of any set of gene differences.

Another way of arriving at much the same place is to consider 'British' history. The British Isles have had the advantage (or disadvantage from other viewpoints!) of being an island group and so slightly less easy to invade, conquer, or otherwise jumble its population up. First peopled by modern humans perhaps 26 000 years ago, the past 3000 years have seen conquest by Romans, attacks by Angles, Saxons, and Vikings, conquest by the French, and, finally, removal of the barrier of the sea by digging a large tunnel under it. All of these events have had a marked effect on 'British' folk.

This history is easily seen by looking at the geographical edges of the British Isles: the present populations of Wales, Scotland, Ireland, and the county of Cornwall in the far south-west of England have some obvious differences from England in language and names, which are Celtic in origin and probably represent the population of the islands some 2000 years ago. Celtic society is quite identifiable and probably originated in the Austrian/German region of Europe and spread throughout northern Europe. The Celts, in their turn, were forced into the extremities of Europe by the rise of the Roman Empire and this is where we still find them today—in parts of Wales, Scotland, Cornwall, and Ireland, where some still speak a Gaelic language as their first tongue (Cornish is by far the least commonly spoken), use many Gaelic place names, and share elements of a Gaelic culture. So we have a purity problem: exactly who are the 'real' 'British'—the Celts or the present inhabitants?

We have, of course, also to worry about those Angles, Saxons, Vikings, and French—for what those foreigners might have done to the British stock does not even bare thinking about! What effect have all those invaders had on the 'real' British? As it happens, it's quite identifiable in some cases. In the far north of Scotland, you start to see names that are not really Gaelic but are obviously more Nordic in origin; even the intonation of the spoken English becomes slightly Scandinavian to a southerner's ears—a sure sign that in the past Vikings stayed more than a night or two. In the south of England, the signs of French influence, following the successful Norman invasion in 1066, are ever present in names and places. The truth is that the British are just like me—a mixed-up population who are no more or less 'pure' than the average mongrel dog.

Is this history reflected in gene differences? It's quite hard to study this obvious question because the historical mixture in Britain is actually a mixture of near relatives—the Celts, Romans, Anglo-Saxons, Vikings, and Normans all derive from European stock that came originally from Africa. So the British are all fundamentally rather similar. Nevertheless it is possible to see some differences in the relative frequency of gene differences between different groups. 'Maps' of the frequency of gene differences within different parts of the British Isles reveal that, even within this small part of the world, this 'local' population has areas where some gene differences are common, other areas where they are rare, and many areas where they are somewhere between rare and common. Put simply, the population of the British Isles has never been a 'pure' population that has subsequently been 'polluted' by non-British genes: it is a rich patchwork of peoples, reflecting its histories and geographies—much as all human populations are.

The future is not so clear

There is, however, a looming change in the dimensions of this problem. At present, we still have relatively little knowledge of the way DNA changes are distributed between different people in any population, but this lack of knowledge is changing rapidly and more and more DNA changes are being discovered in a greater range of populations. The sheer numbers of changes being discovered bring dramatic new possibilities.

Many of these changes will be equally common in all populations; because they are so ubiquitous they are not important and can be ignored. But increasingly DNA changes will be identified that are more common in one population than in others and it is these that will be important as, ultimately, they will provide the basis for classifying an individual's membership of a population.

The way this can be done is best illustrated with a simplifying example. Let's assume that the human population consists of just two groups called A and B— it does not matter, at this stage, what A or B represents. Let's further assume we can find some DNA differences in 90 per cent of A people but just 10 per cent of B people. This means that if we classify people as A by the analysis of a single DNA difference, then we would be wrong 10 per cent of the time; however, if we used two DNA differences, then the chances of being wrong twice are 10 per cent of 10 per cent, which is just 1 per cent—in fact, for every additional DNA difference we use, the chances of being wrong are reduced by 10. It doesn't take long before we have an accurate classification system that will work with a tiny failure rate. Of course, this is a simple example based on a clear population difference of 90 versus 10 per cent, but the important point is that any difference will do—even 51 against 49 per cent; provided there are enough such differences then their final cumulative chance of being wrong will be equally low.

What we have achieved here is not an absolute definition of the group but instead one based on relative chances and I am sure that getting one wrong out of a thousand or a million is probably not a significant error. To put this misclassification rate back into the context of old definitions of 'races', the fundamental division of black and white has historically been based on skin colour. Surprisingly loss of skin pigment is a common genetic disorder that can affect more than 1 in 100 individuals in some tribal groups within Africa. This means that there have always been a substantial number of pale-skinned Africans. Historically, however, the failure of colour to be a strictly accurate predictor of 'race' has not had a significant impact on past attempts at racial classification; a level of misclassification is tolerated even amongst the most racist of ideals.

The most important question posed by the potential use of DNA differences to describe groups is whether the changes are the cause of the differences or the result of them. If DNA changes can be used to describe a human population, could they be the cause of any of the differences between different groups of people? In some respects, this is a question that needs no answer: skin-colour differences between some human groups is obvious, and certainly heavily influenced by gene changes, but there are huge numbers of physical and cultural differences within groups of people that are entirely independent of gene differences. Perhaps the most-precise measure of group identity is language, or the accent of the same language; both are completely uninfluenced by gene differences. The truth, once again, is that gene differences must contribute to some group differences but the extent of the contribution is tiny compared with the consequence of shared culture; I will return to this question in the context of intelligence and group differences in the next chapter.

It is perhaps unfortunate that we are approaching a time where DNA changes will be used to group humans with quite high accuracy and that the patterns of shared DNA differences are likely to reflect a shared history, because this knowledge could provide substantial comfort and ammunition to racist thinking. This is a development that we must watch very carefully if we are to prevent history from, once again, repeating itself in a horrible fashion. The truth is that human beings are different, one from the other, one group from the other, and we all accept, and expect, that this is so. The difficulty with racist thinking is not in the assertion that humans are different, it is in the assertion that our response to individuals should be based on group, not individual, identity; it is the absurdity of this thinking that I will discuss next.

The case of the big woman and the little man

Racist classification of human beings would remain a misguided and wrong view of human origins if it was simply a theory; sadly racist thinking also places a judgement on the groups that the thinking defines—such that one 'race' is inherently superior to the other. This is the primary source of much of the damage and suffering that has grown out of racism. The problem is that racists loose their individual identity and instead assume that of their 'race'. The simplest example is that if black is inferior to white, then all black people must be inferior to all white. As we will see, the absurdity of this argument is extra-ordinary, but it remains the source of almost all forms of discrimination.

Measuring complex features of humans such as behaviour or intelligence is difficult so let's try and discuss group differences with simpler examples, where there are really obvious differences in humans: we can choose height and strength. Let's also try not to get tangled up in the complexities of definitions of populations and relative commonness of gene differences, and instead look at two groups of humans who really do have differences in their genes—men and women. Men and women are both members of the same species but men have genes that women don't have, so this is the racist's dream situation: two types of identifiably different human beings that even a geneticist agrees have real gene differences. Can we use the classification of height and strength as a method for determining who is male and who is female; this is the equivalent of classifying humans into 'races' based on, say, colour or hair texture.

One could imagine that size and strength would be reasonable differences to look at from this viewpoint because men are big and strong, and women are not. Let's expand on this a little: is this statement true? Yes it is; on average, men are taller and more muscular than women. Yes, but is it always true? Are all men bigger and stronger than all women? Of course not! We all know of women who are stronger and taller than many men; the catch is in the phrase 'on average'. What this means is that if we look at enough men and record height and strength, we get a distribution from very tall to very short, very strong to very weak, and there doesn't even have to be a correspondence of short with weak or tall with strong; for example, some tall men are enormous in every respect and immensely strong and there are also tall men who are as thin as a bamboo cane and relatively weak. The same sort of observations can be made for women and we again will see a spectrum of heights and strengths with the average height and average strength lower than that of men.

Let's put all the figures into a single analysis and record the heights and strength of human beings as a whole and now let's ask the sex of the tallest and strongest human. Probably a man. And who is the weakest and shortest? Perhaps a woman. Now the key question: what sex is a person who is moderately tall and moderately strong? We have no idea because it could describe many women and many men. So are all men taller and stronger than all women? Absolutely not! Could I predict the sex of someone from their weight and height? No chance!

The reason for the variation is, of course, that men and women are subject to the effects of gene differences and environment, both of which are busy making big and small humans quite independently of sex. We know that there are real differences, determined by genes, in the way men and women develop and grow, but the outcome is a broad spectrum of height and strengths.

If we now return to the racist view that one 'race' is distinguishable from another by being observably different, we can see that this is an illusion based on looking almost exclusively at the extremes. If we look at skin colour or facial features, for example, then it is not difficult to find Asians who look dramatically different from Africans or northern Europeans. To then conclude that all Asians look similarly different from all Africans or Europeans, as the racial stereotype implies, is as absurd as trying to argue that all men are taller than all women. In practice, skin colour, facial features, and all the other external signs of any one individual fall somewhere within a set of broad and overlapping distributions and these make it impossible to classify a person—we are in exactly the same situation as trying to predict sex by measuring height and strength.

All efforts to find a simple physical way to define racial differences are doomed to failure quite simply because they are all attempting a misclassification of human beings: what any definition of a 'race' represents is an arbitrary classification of human beings. We can illustrate this by inventing an imaginary race of humans who are over 6 feet (1. 8 metres) tall—the race of Giants. By definition, our classification will be accurate: all 6-footers or over would be Giants because this is what the classification determines. The difficulties would start as soon as the Giants had children; not all of them would be over 6 foot because the truth is that height is not a simple product of a gene difference but is instead a complicated mix of contributions from numerous different genes such that Giants can produce non-Giants, and non-Giants can have Giant babies. Even though the initial classification is absolutely accurate in its division of the world into Giants and non-Giants, the passage of just one generation would be sufficient to show that this classification can have no support from genetics.

Groups and differences

Everything I have said so far points to the view that although 'races' of humans do not exist, different groupings of humans most certainly do. There can be no doubt that many of the differences between obvious groups of humans are purely cultural—separate history, language, custom, religion, for example—but this cannot obscure the fact that genes can be distributed differently between groups even though there are no absolute differences; this must be self-evidently true for gene differences that contribute to skin colour. Many examples of gene differences are relatively more common in a subpopulation of human beings and generally this is not the cause of much social difficulty. We are on the whole willing to accept the idea that people with red hair and pale skin are often of Irish or Scottish ancestry and in all likelihood inherited their genes from a distant or not-so-distant Celtic ancestor; at present there is no particular social stigma attached to being of Celtic origin. I would also imagine that to a Roman soldier in Britain some 1900 years ago, a red-haired, pale-skinned individual may well have had a very different social context; I cannot imagine the warring Celts and Romans had the best of social relationships. The important point here is that when perceived gene differences become superimposed on existing historical or religious differences that's when the greatest difficulties emerge.

Perhaps the clearest example of this is the gene difference that is more common in Jewish than in non-Jewish populations: a protein called hexosaminidase A is either different or completely missing in a very unpleasant condition called Tay-Sachs disease. For reasons that are still unclear, this disease is about 100 times more common in Ashkenazi Jews of European origin than in non-Jewish people. Why this should be so could in part be that an Ashkenazi Jew is under a religious imperative to marry another Ashkenazi, and so the gene difference simply became more common by continually being passed around within the population, or it could be that there was some advantage to having less hexosaminidase A protein at some point in the past history of the Ashkenazi—one suggestion is that it contributes to tuberculosis resistance, much as the sickle-cell anaemia gene difference contributes to malaria resistance in some Africans.

It may not be quite clear why Tay-Sachs is so common in Ashkenazis but we do know that Jews in general have a long history of being the subject of racial intolerance and prejudice, and so it is a potentially dangerous, but very easy, step to move from the statement that Tay-Sachs disease is 100 times more common in Ashkenazi Jews to the statement that if you have Tay-Sachs disease, you must be Jewish. The lack of logic to the argument should be quite obvious, but the potential for stigmatization should be equally obvious. An identical case is provided by that of sickle-cell anaemia, where the change is in the globin gene: this is common in people of African origin but not unknown in some Europeans and it is a short step to equating the possession of the sickle-cell gene difference with having 'African blood' with all the prejudice that this can attract. Given the defining history of prejudice and racial intolerance that has surrounded the Jews and Africans, it would be naive to assume that a knowledge of gene differences will inevitably be used in an objective fashion; genetics carries no inbuilt moral or philosophical imperative to beneficent use.

Working on a gene difference that causes disease is obviously important if we are to try and alleviate human suffering, but not all research on genes is of such immediate medical relevance. I mentioned the genetic evidence for our 'out of Africa' origins in Chapter 4 and hope I conveyed some of the excitement I feel as we gain more understanding of human ancestry. A major difficulty in doing research on this topic, however, is that the story of our origins is very rapidly being lost; we are no longer isolated, one population from another. Great cities attract people into a cosmopolitan whirl, jet travel is routine, and marriages cross continental, oceanic, and linguistic boundaries. Gradually the world's population is becoming more and more mixed and, within a just a few more generations, the history of our past, contained in the relative commonness of gene differences, will be gone for ever. The only indigenous human populations that are at present partially immune to this homogenization are those in physical isolation, or that live in groups and select partners from within their own grouping. This means that tribal societies and minority 'ethnic' populations must be prime research material for geneticists trying to tease out our past from the patterns of present-day gene differences.

The work is important: it is not just about our origins, important though this is, it is also about the common gene differences underlying the origin of common diseases, and so there is substantial pressure to carry out this sort of research. If you had to point your finger at just two features of human existence that are major focuses of conflict, however, they would be tribalism and ethnic identity. Genetics has a tremendous potential for abuse, therefore, and it is unsurprising that there can be very substantial resistance to such work from many different perspectives.

The origin of the American Indians (the inhabitants of the Americas before the arrival of European colonizers) has become the focus of much attention and studying gene differences in different tribal populations is an important aspect of this work. Quite recently all indigenous Americans were found to have a difference in DNA—an 'American Indian' gene difference, one might call it, and this was how it was seen in some circles. What went relatively un-remarked upon was the reality of the observation: many non-American Indians also had the change. In short, this was a classic case of a gene difference more common in one population than another, but this fact was lost in the headlines. Given the past history of genocide, the breakup of the Amazonian forest (and associated loss of traditional ways of life), and the hatred aimed at some American Indian tribes by the dominant culture, misuse of this information is potentially explosive.

All of the examples I have used are of gene differences that can be used to stigmatize individuals, probably to their detriment. In many cases, the social consequences might be quite minor, but, of course, within different societies and at different historical periods the consequences might have been much more extreme. I find it inconceivable that the Nazis, had they had the knowledge of gene differences that we have today, would not have used them to identify Jews and I have no doubt they would have had no difficulty in finding geneticists to carry out the tests. The history of genetics in the Nazi era has been chillingly dissected by Benno Muller-Hill in his book *Murderous Science* that makes depressing, and distressing, reading.

A far more pleasant idea is that knowledge of gene differences does not always have to be associated with a detrimental result—indeed, some groups might relish the idea of a simple gene test for group membership.

One drop of blood

Our understanding of gene differences cannot support the idea of exclusive 'races' of human beings but nevertheless questions about 'race' and racial affiliation are still very live issues. Why? Because belonging to a 'race' is a form of identity that many people still accept: many Americans, for example, see themselves as Black, or Chinese, or American Indian. Of course, many of these individuals refer to themselves as being members of an ethnic group—perhaps a less-loaded term than 'race' but equally flawed in genetic terms. The curious problem that crops up with either classification is how to define exactly who is a member of a group or a 'race' and who is a member of a tribe. Historically this question was generally asked by those anxious to preserve the perceived purity of the white 'race' and the discreditable answer has become known as the 'one-drop-of-blood' rule. In the past, this phrase was taken to mean that any distant African ancestry defined an individual as black—the one drop of African blood. In the dominant culture of those times, the contrary question did not seem to be important: is one drop of white blood sufficient to make a person non-black? As the historical role of black and white in our society has changed, this is now an equally significant question.

At the level of gene differences, the concept of 'one drop of blood' is wholly without any scientific meaning. Given we cannot identify a single gene difference that is exclusive to one group, we have no measure of origin for any DNA difference. So I might well have changes in my DNA that more commonly occur in, say, an American Indian tribe than in Europeans: does these mean I have American Indian blood? It could indeed; perhaps my great grandfather George was an American Indian. On the other hand, it is equally possible that I have a distant ancestor shared with the indigenous Americans, perhaps when all modern humans lived in Africa or in the Middle East. Or it could just be chance that I happen to have the same DNA difference. No amount of sophisticated DNA analysis can really help in deciding among these possibilities.

Why do we persist in trying to affiliate with groups of humans? This is often a matter of individual psychology but recently several powerful financial reasons have emerged; for example, in parts of the USA, Native Americans are allowed to organize lotteries and other forms of gambling to raise funds for their tribal activities. Who is a tribal member who could legitimately claim a share of the benefits, and who is not? A very similar sequence of events is unfolding in Australia and I think it is worth discussing this in a little more detail.

Australia, land rights, and ancient peoples

Australia has been inhabited for at least 40 000 years, populated by people who came from islands to the north when the sea level was much lower than it is today and the sea crossings were relatively short. The history of the original Australian people is quite remarkable and it should be read and understood by more of us because it provides an outstanding insight into the origins and histories of all humans.

To a European, Australia seems a young country but the truth is quite the reverse. While modern humans were moving from the Middle East north and west into Europe, the Aboriginal Australians (also, of course, fully modern people, but part of their eastern migration) were changing the face of the Australian continent—altering its ecology irreversibly through hunting and fire-clearing of the brush. The original settlers spread throughout the continent and developed a very sophisticated life that exploited the poor soil and changing weather patterns to the full. At its peak, there were probably some 6 million Aborigines when Australia was finally 'discovered', 40 000 or more years later, by European explorers and colonists. The next few generations saw the Aboriginal population decimated by disease and conflict, and their lands taken over by an alien agriculture and permanent settlements. The Europeans declared that Australia was *terra nullis*—land owned by no one—and so land rights could be obtained over any area that was not subject to a prior European claim. The Aboriginal clans and groups were forced off lands that they had considered as their own, even though they had not fenced them, farmed them, or marked them out. To Europeans, *terra nullis* defined their right to claim whatever they desired.

The effects on Aboriginal society and their social cohesion has been devastating. Most indicators of poverty—high infant mortality, high suicide, high alcoholism, low income, and poor housing—show that the affluence and delightful lifestyle of much of the rest of Australia has, in the main, passed by the survivors. Until quite recently there seemed little prospect of any of this changing, but gradually there has been a courageous attempt to look at the recent history of the treatment of the Aborigines and accept that the concept of *terra nullis* stemmed from a very blinkered view of the continent's history. In a series of landmark court cases that are still continuing, Aboriginal land rights are being accepted as valid and are being granted. The revisiting of ownership produces some wonderfully juxtapositions of old and new cultures; my favourite was when the local Aboriginal peoples set out to reclaim the Gold Coast—a very popular hotel and tourist destination in southern Queensland with numerous expensive high-rise hotels, casinos, shopping malls, restaurants, and all of the trappings of mass tourism of the most successful and materialistic kind. I do not wish to make light of the underlying issues but the contrast between old and new could not be more marked!

The key question that now clearly emerges from all of this is, once again, how does one define a member of a tribe? The tribal lands that are being reclaimed by the indigenous Australians include not just the Gold Coast but many other prime pieces of real estate, and some lucrative deals will have to be worked out as to who pays whom rent—the tribes will certainly become richer if not rich. Given that the Aborigines generally left no written records, the question of who is a tribal member is not easy, and the whole area of gene research within these peoples with their ancient history has suddenly assumed a quite new dimension.

The scientific study of Aborigines has sadly had a tawdry history, all too often characterized by the treatment of Aborigines as being somewhat more of a 'specimen' than a fellow human being—a specimen of 'primitive' humanity that was to be exhibited in a museum case. Australian Aborigines are one of the oldest isolated human populations with a dramatic ancient history. Studies of gene differences show clearly that they originated in the islands to the north of Australia but they also show that, like in most peoples of the world, Aboriginal gene differences are being mixed with the new European and Asian gene differences, giving rise to potential areas of difficulty as far as the uses of information derived from genetics is concerned. Nevertheless, the research can show whether it is possible that an individual has Aboriginal ancestry.

In either case, this information can be used to stigmatize or exclude an individual and it is therefore imperative that research is carried out with the utmost sensitivity to these dangers. In particular, there is a need to obtain consent to the experiments based on a detailed understanding of goals and limitations of the derived knowledge. The need to explain and educate cannot be over-emphasized: the sensitivities of several hundred years of discrimination have left a great legacy of distrust and I do not believe that geneticists have any right to ignore history simply because they are scientists—the goals never excuse the means. In medicine, 'informed consent' is central to all forms of medical treatment and there is an absolute requirement that the benefits and dangers of a treatment must be explained (the 'informed' part) before consent is obtained.

Informed consent is also a concept that can, albeit with greater difficulty, be sought from groups. A group is, of course, a collection of individuals and there can be quite different needs even within a small group, so this process can be difficult and divisive, but certainly valuable. In practice, scientists have to go to great lengths to explain, educate, and get permission to carry out work on groups: entire research papers have been written on this single topic. For example, one of them details how permission was sought through Apache tribal elders, who called meetings with the community to discuss the benefits and dangers of genetic work that was proposed and the researchers had personally to explain why their work was important and what benefits might accrue to the community.

The distribution of gene differences within human populations, then, cannot be used to support the existence of 'races' nor prove the group membership of an individual. Anybody who wishes to believe otherwise does not understand the dynamics of gene differences that I have described. Gene differences are, however, clearly a reflection of human ancestry and so can be used to support some aspects of group or tribal identity by confirming a shared history. There is at least one case where they have been successfully used to do just that—in the case of the black Jews of southern Africa, the Lemba.

The Black Jews of southern Africa

One of the more remarkable cases of using genes to try and establish membership of a group is represented by the studies of the Lemba people of southern
Africa. The 40 000 or so Lemba live mainly in Zimbabwe and northern South
Africa and are dark-skinned, but seem to have some features that are more
non-African than might be expected of other African people. Their language is
different from that spoken by most of their neighbours and is unintelligible to
the surrounding Bantu speakers. Culturally they are known for their skills in
metal work and their social customs strongly discourage marriage outside
the group, which is also held together by a religion with very unusual ritual
and laws. The food laws seem remarkably similar to those of the Jews and this
is the really unusual feature of the Lemba. Their oral tradition states that they
are of Jewish origin and came into Africa some 2000 years ago as traders and
metal workers; some stayed in Ethiopia and are the modern day Falashas—the
more widely known 'Black Jews'—but the Lemba's ancestors undertook a long
journey south.

Could a study of DNA differences support the origin of the Lemba? DNA
testing for differences of various types showed two interesting things. As
expected, there are some DNA differences that are more common within the
Lemba than neighbouring peoples, as might have been expected from their
tribal organization and selection of partners predominantly from their own
group. What is more remarkable, though, is that the pattern of commonness
and rarity of gene differences is most similar to the pattern in Jewish people (the
technical term is Semitic): the Lemba do indeed appear to be more similar to
Jews than to Africans.

Very roughly, the analysis found that about 50 per cent of the gene differences
have a Semitic origin, about 40 per cent come from African neighbours, and the
remaining 10 per cent from some unknown origin: this seems to be clear support
for a Jewish origin of the Lemba. Curiously, though, the Lemba DNA differences
do not seem to be that similar to the Falashas—the Ethiopian Jews to which
the Lembas claimed to be related. It isn't yet clear if this is simply a technical
problem—too few samples analysed, for example—or a biological effect of
marrying within the group, which is known to have effects on commonness
of DNA differences, or if this part of the story is historically inaccurate.

The sharing of DNA differences does not necessarily 'prove' the Lembas are
indeed a displaced tribe of Jews, because similarity in patterns of rare and
common gene differences can happen by chance and all we can really conclude

is that the history contained in their genes is not incompatible with a Jewish origin. The phrase 'not incompatible' sounds terribly weak but it is an inevitable consequence of the lack of differences that are unique to single human populations; there must always be a chance that the similarity is the result of independent events and not shared history. As usual, the proper position for the geneticist is to sit on the fence and suggest that more work on the Lemba (meaning analysis of more gene differences) is needed before the relationship with Jews is firmly established. Even as I was writing this, the results of tests on more Lemba individuals using different DNA differences have confirmed the original observations and we can conclude with some measure of certainty that the Lemba really do appear to have a shared history with the Jews.

It is a wonderful story: modern genetics has proved what the Lemba knew all along. But it might not have ended this way: what if the DNA differences had shown the Lemba were not likely to be of Semitic origin? What effect would this have on the self-image of the Lemba? Lemba culture, like the culture of many relatively small groups, is under continuous pressure as they become exposed and assimilated into the culture of their more numerous neighbours. Perhaps the knowledge that science had shown that their origins could not have been as their elders taught them would be sufficient to destroy the very thing, the Lemba culture, that made them different and cohesive. Would a geneticist have a right to risk this outcome? Clearly a study like this could have an enormously harmful impact on a group if it was handled insensitively. In this case, therefore, all the samples were obtained from volunteers who were attending the Lemba Cultural Association annual festival, where the research aims and outcomes could be properly discussed.

If there are no 'races', what are ethnic groups?

From the genes' point of view, all humanity may look rather similar but all of us are aware of the contribution that culture and history have made to human societies. We choose to call the resulting groups 'ethnic' rather than 'races', and genetics seems to recognize that ethnic groups are real entities when looked at from the genes' perspective. Why do I say this? Because geneticists often record and discuss the 'ethnicity' of individuals studied in any genetic research programme. Surely this is tacit recognition that ethnic groups must have a scientifically supportable existence even if 'races' do not? Remarkably, despite all the uses we make of ethnic groups in genetic research, from the genes' viewpoint individual ethnic groups, just as is the case for 'races', have no uniform genetic identity.

If this is so, why do geneticists study ethnic groups? To understand this, we need to think a little about how scientists do some of their work; the testing of new drugs provides a good example. When researchers test a new drug on human beings, they are primarily seeking to show that the drug is effective and that it does not have dangerous side-effects. One way this is done is quite simple and involves splitting a group of people who all suffer from some condition into two equal halves. One group (the 'cases', in the jargon) are given the drug; the second group (the 'controls') are not, and so in principle the cases should all respond favourably to the drug and not show any side-effects. The 'controls' might seem superfluous but this is far from correct because the cases, by definition, are not well. Are any symptoms that they show simply a mani-festation of the disease, or are they caused by the drug? If similar new symptoms appear in the controls, then this question is simply answered: it must be the disease because they have never been given the drug. This seems a very straight-forward and clear experiment but unfortunately hidden within it are some difficulties that are best understood by inventing an imaginary drug trial on a hypothetical new drug, Aixrgone, which is meant magically to reduce pain and nausea caused by drinking too much alcohol.

To show our new drug works what we have to do is to feed large amounts of alcohol into a group of unfortunate test subjects so that they end up hopelessly drunk and, later still, terminally hung over. We will then treat half with Aixrgone and half with something that appears similar to the taste and eye but contains no drug and is useless. It is essential that we researchers do not know who gets Aixrgone and who the useless pill, in case we are biased in our obser-vation of the drug's effects. After a period to allow the drug to work, we carry out a battery of standard tests to see if Aixrgone has alleviated the pain of the hangover; only later, when all the results have been analysed, will the identity of who got the drug and who did not will be revealed. If Aixrgone works, a clear relationship will be established between lack of pain and taking the Aixrgone, but not the drug-free, pill.

Case/control studies are an excellent method for rigorously testing many aspects of human biology, not just drug-testing, but they are unbelievably easy to get wrong. How? Imagine our Aixrgone test, but this time the people we are going to test are a mix of Japanese and English. About 50 per cent of Japanese (and many other Asiatic nationalities) cannot make normal amounts of a pro-tein called aldehyde dehydrogenase 2, which is involved in breaking down alcohol; this is the reason why many Japanese rapidly develop a red flush in their face when they drink alcohol. The condition is almost unknown in English people. If we have Japanese and English people in our Aixrgone study, is it not a reasonable proposition that the Japanese and English may respond to Aixrgone quite differently simply because they break down alcohol at very different rates; the Japanese might, if given the same amount of alcohol as the English, suffer much worse symptoms than the English and so might report that Aixrgone did not work for them. In this study, therefore, it's important that English and Japanese are equally represented in both cases and controls, for if the Japanese were all the cases and the English all the controls we might erroneously conclude that Aixrgone was ineffective.

The illustration I have used is trivial, but to overcome this sort of problem real case/control studies have to be constructed very carefully. For example, even though the large tests conducted by the pharmaceutical companies are always done in many countries, it remains critical that cases and controls are as genetically similar as possible. This is achieved by selecting both cases and controls from the same ethnic or national group—Japanese, English, Arab, Scottish Highlanders, Indians, Africans, and so on—because by doing this, it should be possible to reduce the chance of the ethnic differences affecting our understanding, just as it might have done in our Aixrgone tests.

The use of ethnic classification suggests that geneticists preach one thing but practise another: on the one hand, they use the existence of a continuous spectrum of gene differences across human populations as the main argument against the existence of 'races' and, on the other, they deliberately select ethnic groups to minimize the genetic differences between cases and controls, suggesting that ethnic groups must be genetically homogenous.

The reasons for this apparent double standard are simple enough: at present we really do not have any easy way of determining the 'genetic relatedness' of different human beings. From the scientific viewpoint what we want are people in case and control groups who have, as far as practical, the same set of gene differences; we would not even care if they had the same ethnic background provided this condition was met. At present, we have no way of assessing genetic relatedness directly and so we are forced to use ethnic groups as a crude way of reducing, but not eliminating, gross differences in gene differences in the test group. In future, this will change: I will no longer be a male Caucasian but will instead by a male of variation type XYZ, where XYZ will be a statement of the key gene differences I contain. This means that scientists will be able to identify groups of humans with the appropriate similar or identical sets of gene differences and use them as the subject of case/control studies, in the certainty that there really is a genetic match across all individuals. At this time, ethnic groups will cease to be a term used in genetic description and it can revert to its accurate use—groups of humans defined by common cultural and historical links.

Ethnic weapons

We can be certain that we have not heard the last word on genes, 'races', populations, and groups. Our understanding of the variation of human DNA is progressing at a precipitous rate and great attempts are being made to apply this knowledge to human populations in the search for the common differences that contribute to medical disorders. It is important that geneticists recognize that this research will in all likelihood provide the means of establishing much more robust boundaries to populations than they are currently able to do, and it is important that they are prepared for the social consequences of this eventuality.

The potential of gene differences that are limited to a single population has already opened up uninformed discussion of potentially totally new ways of exploiting group differences, of which the most ominous and headline-catching is the possibility of creating 'ethnic' weapons—drugs, chemicals, or infectious organisms that are 'targeted' specifically to affect only members of a chosen and hated group of people (you can choose your most likely target). If scientists come to define differences within genes that are specific to a single group of people—remember, such changes have not yet been identified—then it may be possible to develop methods to 'target' such people specifically.

I think that this is an unlikely possibility for three good reasons. Most 'hated' groups are usually quite closely related to the hating group and often have interbred because they frequently share common territories; this means we are very unlikely to find usable gene differences in such closely related groups. The second major problem is even if we were able find such a difference, and even if the protein would represent a viable weapon target (an extremely difficult problem as numerous attempts to make biological warfare weapons attest), there are still huge biological problems in using this as a target for a weapon. Primary amongst these is the difficulty of making the weapon interact only with the appropriate difference in the protein target—this, after all, is the basis of the selectivity of the weapon. Interactions with proteins are rarely completely specific, which means that any chemical weapon will be at best 'relatively' specific in its interaction with the ethnic 'specific' form of the protein. This in turn means that the weapon will have only a relative specificity; it will probably kill many of the hated group but also end up killing a 'relatively' small proportion of your own population! Even for the average crazed dictator, such a path will be quite difficult to justify—few populations will tolerate being relatively dead! Every step of the path is unlikely, and my belief, therefore, is that ethnic weapons cannot be made and are science fiction. I have never been a great believer in the overall level of human intelligence, however, and there have been, and no doubt will continue to be, shadowy groups pouring over the maps of human gene differences; they will remain an irrelevance to all but authors of sensational news headlines and ill-educated despots.

In conclusion

To summarize, our present understanding of genetics firmly refutes ideas of purity and exclusivity, and shows that classifying people into groups based on rigid differences is arbitrary and has no basis in science or history. As I hope I have shown, genetic knowledge remains rather limited and, in particular, understanding of the variation within human DNA is as yet rather fragmentary. In time, the 11 million or so common variations in our DNA will be identified and then it will become possible to seek better and better genetic indicators of group membership by using more and more DNA differences to define them. Does this mean it is just a matter of time before geneticists stumble on the key to unlocking 'race' membership? I suspect the answer is in part 'yes' and in part 'no'. My guess is that the differing ages of human populations may make it easier to define groups with a recent historical origin but that the most ancient groups will remain indefinable, But even this degree of uncertainty will provide new ammunition for racialist thinking.

Am I implying that gene science can never kill racial thinking? It must be clear that differences between human groups do occur, even if these differences are relative rather than absolute. Ultimately racial views do not stem from an understanding of biology and genes and I suspect that if we are to understand 'race' and racialism we probably need to look much closer at human behaviour and human psychology as it relates to group living and activity.

I have discussed 'races' and ethnic groups at some length and my conclusion is that neither has a supportable existence based on simple gene differences. But I am sure that none of us believes that genetics has lain to rest the whole problem. 'Race' and ethnic identity are cultural concepts defined by history and social practice born out of shared experience, and genes cannot be used to disprove the existence of history; geneticists must continue to struggle to prevent their science from being used to support untenable and discriminatory definitions of group identity.

'Race' is not the only battleground within gene science; huge battles have also been fought over how gene differences might influence our minds and our intelligence. These controversies have themselves fed back into arguments about racial differences and I will return to this in Chapter 8. But before we can do this we have to understand how genes contribute to the most complex and perhaps least-understood organ in our bodies—our brain.

Genes and the mind

6

We can now begin to understand why Jeanne and Jean are physically so different—because both their gene differences and their environments have made them so. We could easily have spent a great deal of time discussing what is understood about how genes control our development and how this might be modulated by our environment and our gene differences, but this would really only tell us what we already know: that genes are important in defining our body plan. The far more interesting question to discuss is how far the gene's influence might extend? If the body is shaped by gene differences and environment, what is the effect of genes on the brain and its workings? How much of the mental abilities of Jean and Jeanne were determined by their environment and how much might have been conditioned by their gene differences?

The answers to these sorts of question are often very disturbing to us. All of us like to feel that we are in control of our own actions and so the possibility that our response to events might be conditioned by our genes, quite outside of our control, is viewed with great dismay. The truth about genes and our brains should be reasonably heartening to anyone who feels like this: genes affect us but in a way that is far from predictable and, furthermore, only very rarely does our behaviour become irreversibly affected by the genes within us. What we are going to see in this chapter is that the effect of genes on the brain is similar to the effects of genes on our bodies—that genes define many of the most fundamental features of our brains; armed with this knowledge we can then move on and in the next chapter I will discuss how deep this influence might extend.

Once again genes are touching the separate lives of Jeanne Dream and Jean Battler, but this is not the only influence at work.

■ A study based on ignorance

When we look at the influence of genes on our brains the difficulty is that we have no detailed idea about how the brain works. This is unsurprising because the brain is staggeringly complicated: it contains about 10 000 million nerve cells and each nerve cell can be in electrical contact with about a thousand other cells, making a staggering 10 trillion (million million) connections! Each nerve cell makes electric contact by physically touching another cell and it can do this because it has a central 'body', where most of the cell protein and chemicals are made and where its DNA is located, and from this body extremely long, slender outgrowths extend to touch other cells. These outgrowths can stretch past many cells in any direction and so it is unsurprising that sorting out exactly which cells are in contact in a human brain is a project that is way beyond our reach.

The next problem we face is that we really do not understand how all of these connections can operate together to make the brain undertake its many and complicated roles. We know that electrical connections are central to its functions but quite how the brain can store information or control complicated tasks is still not understood. It is tempting to fall back on the usual analogy—a human brain is like a computer, only more complicated. But this really doesn't work as a useful analogy because a computer stores information in a very simple way: each part of its memory can be 'on' or 'off', like a light switch. This means that each bit of memory needs to have a very limited number of connections to record complicated information as a series of on/off codes—generally a couple of connections per on/off memory. In contrast, the basic memory unit in a brain is probably a single nerve cell with its 1000 connections—hugely more complicated than a computer memory. So the computer/brain analogy is a poor one, which puts us in a difficult position: if we do not know, even at the most basic of levels, how the brain works, how can we see if genes influence its workings?

This problem is bad enough on its own, but the brain is a very unusual organ in the way it responds to the outside world and this makes life even more difficult for gene researchers. Why should this be so? Let us consider what a brain has to do during its early life: it has to 'learn' a multitude of critically important tasks—how to acquire and manipulate speech and grammar, how to recognize individuals, how to control movement, to name but a few. All of this has to be learnt in early childhood when the brain is very different from what it will be in the adult.

It is tempting to believe that the reason a baby cannot talk is simply that it has not learnt; this must, of course, be partly true but it is only an element of the problem. The real difficulty is that a baby's brain is not properly wired up and so babies simply cannot do some of the things adults can perform effortlessly. Why should this happen? Because the brain is like no other computer or anything humans have ever made: it is an organ that uses information from the outside world to develop its own internal electrical connections. A baby cannot see as well as an adult because some of the 'seeing' wiring is not connected, and the act of a baby looking helps make the wiring connect up in the right way. In short, the environment—what the baby sees—has a great influence on the final structure of the seeing part of the brain. This is a very subtle idea: no machine designed by a human can assemble itself in this way and the brain's ability to achieve this is a wonderful piece of organization. But it is also the geneticist's nightmare: how are we ever going to be able to separate the effects of genes and environment if the environment in part builds the brain? It is this, even more than the enormous size of the problem, that makes brain research so difficult.

Can we escape from these difficulties? Is there a different sort of approach that we can use that may enable us to identify whether there is role for genes in the brain? Probably yes there is; and, most importantly, it does not require us to have a real understanding of the detail of brain function. It relates back to the human body plan. Even though we do not know the full detail of how genes make the body plan, we know enough to know that they do it. Could we show that genes are just as important in forming the 'brain plan' as they are the body plan?

A body plan of the human mind—yes, by genes alone?

What would a brain plan look like, given we really know little about how the brain is wired up in either a newborn baby or an adult? Perhaps two predictions are possible. First, the brain should have a 'structure', which is broadly similar in all people (much as our body plan is similar in all of us). The brain could be said to have a structure if we could show that specific regions always have the same function in different people, irrespective of the 'environmental' effects of sensory experience. Are the same regions of the brain always used to enable different humans to see, or to hear, or to talk? Second, the brain should give us abilities that are common to all human beings and that are not dependent on the way we are brought up or other influences of environment. A good example of what I mean would be that virtually all humans can talk, suggesting that our brains are designed to enable us to speak, even though the actual language we learn depends heavily on the cultures we are exposed to.

Why are these important predictions? Because if a brain has structure that is independent of experience, then the only way we know such complex structures can develop in humans is through a programme controlled by genes, a genetic programme (much like that which defines the body plan). Similarly if we can see universal behaviours in humans, then this would suggest (but not prove) that there is an underlying similarity in brain function that may be the functional expression of the brain plan. Is there evidence for either of these predictions?

The human brain in modules?

There are two extreme versions of how a human brain may grow and wire itself. In one case—the 'genes do not control' possibility—the brain might organize itself entirely through nerve connections that somehow become refined and fixed by the effects of the electrical signals coming in from the outside world. Imagine that in its earliest stages the brain was simply a mass of nerve cells that may or may not have made connections between themselves; electric signals coming in from the eye, carried by a nerve that links brain and eye (the optic nerve), could influence the nerve cells in the brain and this influence could make or break or reinforce connections. Gradually some region of the brain would start to develop all of the functions that are needed to enable us to see and understand what we see. The idea is attractive because it would mean that the brain does not have to have detailed plans of how it is to be wired up—join nerve 10 000 to nerve 11 000, and so on; which would mean an incredibly complicated set of instructions. Instead, wiring up would occur much more spontaneously.

The alternative, the 'genes control' possibility, is that genes control all aspects of a nerve cell's life by using proteins to specify every part of the process, and hence the wiring of the brain is completely defined by their activity.

These are the two extreme processes that we can put forward; there is no reason why genes have to control all aspects of the brain to the same extent. The end result of both of the two processes would be a functioning brain, but there is an important distinction. In the case of the 'no genes' possibility, one might expect different brains to be organized in different ways, even if they all have essentially the same functions and abilities. In the 'genes only' case, the prediction would be that every brain would have an identical organization, much as our bodies have identical organization. So if we could show that every human brain had the same overall organization—for example, that the same region controlled sight in everybody—then this would strongly support the 'genes control' idea. In contrast, if different people used different regions of their brains to do the same thing, this would be good support for the 'genes do not control' view.

In reality, there is a large amount of evidence that the brain is organized in very specific ways and that the same region is involved in controlling sight or speech in all of us, which means that the 'genes control' view is likely to be the more important process.

What is the evidence for this? Many different threads can be drawn together to show this and I'll discuss two of the most dramatic only. In a rather direct and unpleasant way wars and, more recently, car accidents have taught us a great deal about the way the brain is constructed. If the brain is organized into discrete areas that control particular parts of our behaviour, then it follows that if we remove or damage the same area in different individuals their behaviour will be affected in the same way. Wars and car crashes do this experiment for us by generating large numbers of individuals with damaged brains. The observations are clear: damage to the same part of the brain in different individuals tends to change the same parts of their behaviour. So, as long ago as the 1860s, it was discovered that damage to two areas of the front part of the brain results in a profound affect on speech. Damage to one of the areas resulted in an inability to speak and, to the other, an equally deep inability to understand speech. The two brain regions—now known as Broca's and Wernicke's areas, respectively (named after the surgeon and physician who were involved in early research in identifying them)—seem to be very important in controlling how our brains function in these quite specific ways.

Since the nineteenth century, many other examples of damage to specific brain regions that causes impairment to specific human behaviours have been identified. There is now general acceptance that the brain is almost certainly composed of a series of regions or 'modules' that have particular importance in controlling different aspects of our brain function. This is not to say that the 'module' is all that is needed to make us talk—many other areas of the brain can contribute as well.

There is a large body of more modern experimental information supporting the existence of modularity within the brain, much of it derived from the effects of strokes and of surgery (the latter sometimes involving quite deliberate destruction of parts of the brain). But all of this type of information suffers from the same general problem—that we are looking at the behaviour of the damaged brain and not the normal. Perhaps modularity is a peculiar and unnatural response to damage itself? This seems to be unlikely—how on earth could the brain reorganize itself so radically? Fortunately direct evidence for modules has recently been provided by brain researchers, who have taken pictures of the working brain using some very sophisticated new viewing methods.

A snapshot of the brain at work?

Imagine we had a camera that, instead of taking conventional pictures of a human body, could look into the body and take pictures. Of course we have X-ray machines, which can look through soft tissues and bone, but imagine we could take a picture of the types of chemical in a body rather than just soft or hard regions. This chemical camera exists; instead of X-rays, we can use magnet fields to 'image' different chemicals in the body and, not only that, we can produce images that are so detailed that we can see inside the organs of the body and down to objects just a few millimetres or less in size. These cameras take pictures using a process called magnetic resonance imaging, or MRI for short. Fortunately how the method works is not important to understand—it is complicated physics.

There are several different ways MRI cameras can be used. They are not similar to the cameras used to take pictures; not only do they not look similar, they 'see' very different things. The most important machines, from our point of view, are the ones that 'see' changes to blood flow in the brain. When the brain works, it does so by sending electric currents down appropriate sets of nerve cells; these currents have to be made by the nerve cells and this requires that the cells expend energy—much as energy is expended to make a generator produce electricity. In the case of a cell, energy is stored in a chemical form and 'burnt' using, amongst other things, oxygen. If a cell is to produce electricity it needs more oxygen as well as other chemicals, all of which it gets from the blood. So when an area of the brain starts to work, two things happen: first, blood flow is increased to bring in more oxygen and chemicals and, second, the amount of oxygen in the blood leaving the brain is much reduced compared with the amount when the nerve cells are not working. Remarkably it is the difference in blood flow and oxygen content that these cameras 'see'.

It is important to realize that the camera is not 'seeing' the electrical activity of the brain in a direct way but is rather 'seeing' the aftermath of this activity. In particular, the changes in blood flow and oxygen levels can alter in complex ways that do not have to mirror either the exact timing or the amount of electrical activity; these are important reservations that need to be kept firmly in mind when we discuss what it is that these machines tell us about our brains

The experiments that can be done with MRI are all rather similar; the experimenters simply attach the machine to a person's skull and get the individual to carry out a series of sometimes quite complicated tasks. The MRI cameras will detect any areas of the brain where the blood supply is altered, indicating that they are involved in more electrical activity than other regions. It is then simply a matter of the experimenters observing the brain before, during, and after some specific set of tasks and this will allow them to identify any region of the brain that is particularly active while the task is being performed.

MRI pictures often show clearly that different regions of the brain become active when different tasks are performed and, most importantly, the same regions of the brain always seem to respond in different individuals when asked to carry out the same task. This seems to be strong evidence for modularity in the undisturbed brain (MRI, by the way, does not seem to have any influence on the brain's function).

My favourite experiment using MRI not only shows the modularity of brain activity very convincingly but also shows why environment is so important to the brain's function. If you use an MRI machine to observe people talking, then one area that seems to be involved is Broca's area, just as you might have predicted. It is important not to oversimplify this experiment because many other regions of the brain also seem to be involved, and these are clearly detected in the MRI pictures. Perhaps, then, Broca's area is of central importance but other regions also process the information needed to talk.

In part, the MRI pictures from this experiment are simply modern technology confirming what we already knew, but they were much more interesting than this because the scientists studied two groups of people who could talk two languages. The first group had learnt two languages from birth—they were brought up as bilingual, using both languages with equal ease. In contrast, the second group was brought up using just one language and learnt the second language only later in life. The experimenters then asked the people from each group to talk in first one language and then the second, whilst they took MRI pictures. Quite stunningly the result was that those speaking the two languages from birth used pretty much the same region of the brain when talking either language, but those who had learnt the second language late used two close but separable regions depending on the language they were speaking. I find this result enormously satisfying because it illustrates a very important point: the brain may be modular but it can change dramatically depending on the sort of input—its environment, if you will—to which it is exposed. In this respect, the brain, as I stressed earlier, is unique in our bodies.

The evidence for a structure for the brain is overwhelming even if the structure can be heavily influenced by the brain's own activities. How do genes fit into this picture? As I made clear at the beginning of this chapter, we really do not know in any detail. All we can say here is that a brain has a structure and that this is most simply explained as being the result of the same sort of processes as give our bodies a structure. Do we understand how genes define the development of the brain?

The human brain has evolved over many tens of millions of years from a much-simpler organ in our animal ancestors and so many of the features of, for instance, the Hox code (discussed in Chapter 2) have become obscured by the complicated folding and refolding of the brain's tissues. The rear part of the brain, called the hindbrain, is rather simpler and in this region there is clear evidence that the Hox proteins are important in determining its structure. But to extend this understanding into the parts of the brain that control our more-complex skills—the forebrain in particular—will require much more research. So for the moment, I think we can conclude that the brain's structure is perfectly compatible with there being a brain plan, with this controlled by genes. The next question must be, if the brain has a plan do all human brains have the same set of fundamental activities?

Do all humans have the same abilities?

It is obvious that most human beings have different skills. Indeed, we all recognize that our friends have different personalities and different abilities. But, what is underlying these abilities? Let us go to what is, in one sense, a trivial observation. No one has any difficulty in recognizing that we are dealing with a human intelligence when we talk to other people. I do not mean that we can see a human in front of us and so know we are talking to one; I mean that we would never have any difficulty in deciding whether we were dealing with the mental abilities of a human as opposed, say, to a cat, a dog, or even a gorilla. We expect human beings to be able to do certain tasks that are the products of our brains. We expect a person to talk, to be able to see, hear, think, respond, laugh, cry, and reason. Some of these features are shared with animals, albeit developed to a lesser extent, but we know what we expect of a human being, even if we cannot be too sure of exactly where human skills start and animal skills cease.

In short, we all know, even if we might not be able comprehensively to describe it, exactly what we expect of another human being. This seems to me to be quite important: it suggests that the human brain is not unlimited in its function—on the contrary, it is rather specific—and this is the second reason why there is probably a brain plan. It looks as though the brain is designed to carry out some rather universal tasks and is not simply an unstructured recording machine that is forced down particular activities by our sensory experiences.

Can we really be sure of this? After all, all human cultures and societies seem to have language and so perhaps all brains simply learn a language because they are exposed to it. The best evidence that this is not true comes from the work of both brain researchers and also of linguists—people who study languages. Perhaps foremost amongst linguists is Noam Chomsky, who proposed that the ability to learn a language in humans is 'innate'; by which he meant not that our brains were built to talk English or Russian but that they were specifically organized to learn a language. Not only that, the reason we could learn whatever language we were exposed to from birth was because all languages have a common structure within them, a common grammar. This sounds complicated but at its simplest it means that human brains are designed to talk a language with a certain structure to it. We are all familiar (if rusty, in many cases!) with rules of grammar—nouns, verbs, adverbs, and so forth. Many languages have different rules of grammar: for example, German requires verbs at the end of sentences whereas English does not. Chomsky suggested that the grammar of all languages (and there are about 4000 of them in the world, all with rules of grammar) could be underpinned by a universal grammar—what, very roughly, the brain contains. This is why a human baby can learn any language, because its brain is designed to do just that.

The rules of this grammar are gradually falling into place but, as you might imagine, working on 4000 or so languages is neither easy nor without disagreements. I do not for a second want to give the impression that this is an area of linguistics with which geneticists are concerned: what genes have to do with universal grammar is totally unknown. But the existence of universal features of even apparently massively complicated and uniquely human attributes such as language means that the brain is perhaps a great deal more influential than we readily accept.

As I have said, we do not have the faintest idea as to how genes could define brain structures that contain a universal grammar. But the existence of such grammar seems the very best argument that genes are responsible for making the brain structures that in turn hold the rules of such grammar and that also implement the rules when we speak or understand language. How else could the brain develop such abilities except under the influence of genes? Are there other possibilities? One alternative would be that the universal grammar rules are the rules of the first language spoken by humans and so underpin all languages spoken today. This rather begs the question of how humans came to acquire language in the first place and why our primate relatives failed to develop it.

But I think that the universal grammar is probably a product of the connection and growth of nerve cells in the appropriate modules and regions of our brain, and the fundamental organization of these regions is determined by proteins encoded by genes. To think otherwise is to make the brain an impossibly special place in the human body and ignores the universal features of the human mind. If we think of the gene's influence on the body plan and the environment's influence on body shape, then why should the brain be different? The only question that remains is just how much of our behaviour is influenced by genes? Is it just the brain plan or do genes touch on personality or behaviour as well?

■ The gene's influence on the mind?

Exploring the influence of genes on the human mind is a subject fraught with difficulty. We understand so little about how the mind works that we are forever having to measure something that is an indirect product of the mind rather than any more specific measure of the way our minds actually work. The analogy that I think best shows why this is a problem is to think what would happen if an alien landed on our planet with instructions to 'understand cars'. What would the alien first study? In the absence of any prior knowledge, the hapless alien would perhaps decide that the colour of cars was the most important feature of their function and classify cars as red or blue. There is nothing wrong with this view; it simply misses the point that cars have a myriad of different features that can be used to classify them—engine size, petrol consumption, tyre width, power, and so on. In short, if the alien understood how cars worked, and their use, then classification would be easy.

This is the position that we are in with the study of the mind: we can look only at general properties that may be difficult to measure because we simply do not know how our mind works. There is also the additional difficulty that our mind has many different skills—reasoning, language, perception, and personality. How do we 'measure' such diverse skills; what is the measure of the mind? If we unable to measure the features of a mind easily, doesn't this imply we will find it impossible to do any of the studies to detect influences of genes on the workings of the mind? This is probably the key difficulty that underpins a great deal of the argument that rages between genetic scientists that do, and those that do not, believe that there are influences of genes on the mind.

So what we will do is first consider some conditions that show the genes can have a massive influence on the mind—an influence so great that there can be no argument because the mind is so obviously affected. Then we can move on to conditions where measuring the extent of functional disorder in the mind is more difficult. Finally we can turn to the most difficult problem of all—to look at the proposition that gene differences influence the workings of the normal mind.

Rare differences in proteins alter behaviour

Brain cells make more different proteins than almost any other cell in our bodies—about half of all genes are switched on, presumably reflecting the great complexity of the brain's structure. Genes must have a central role in controlling the production of these proteins within every nerve cell and also in controlling brain development; we can imagine the details of this control being very complicated simply because the outcome is a structure of huge complexity. Earlier (Chapter 2) I argued that the body plan was invariant because of the nature of the hierarchy of gene regulators, and this suggests that the brain ought to be similarly difficult to change; this is indeed the case—we all have the same brain plan. So it should be surprising to discover that the brain can be dramatically influenced by differences in genes; indeed, several genes seem to be able to do exactly this. How can this be so? Is the brain plan different from the body plan? No: in these cases it is not the brain plan that is changed, it is the most complicated of the brain's finished abilities that is altered—the brain's intelligence and its control of behaviour.

There are several hundred different conditions—most of them, fortunately, very rare but others remarkably common—that cause changes to the behaviour and abilities of humans and for which there is good evidence that there is an underlying influence of genes. Knowing which genes are involved in these conditions is of tremendous potential importance to us because it opens immense possibilities for treatment and diagnosis and even to understanding how the human brain works.

Measuring the mind

Before we go much further we have to face one major problem. I have blithely referred to genes that alter human intelligence, but this assumes that there are simple ways of measuring intelligence in the first place. It's possible to reduce the intelligence of a person—accidents causing brain damage are the most obvious, and distressing, way of achieving this. And we certainly recognize the mental damage in an individual whose behaviour and skills were known to us before and after some catastrophic accident, simply because we can compare the before with the after. In contrast to accidental damage, some people are born with mental difficulties. Such 'mental retardation', as it is called, causes more difficulty for the simple reason that there is no 'before' and 'after' to allow us to compare mental abilities.

How, then, can we recognize mental retardation in a person born with some brain disorder? There are two obvious different general measures: intelligence and ability to cope in life. Why these two? Intelligence is perhaps self-evident but begs the question of how we measure intelligence. Ability to cope is an interesting measure because it reflects many different features of intellectual ability. It relates to intelligence: if you cannot understand the concept of money and numbers, it is difficult to shop unaided; if you cannot comprehend a reasonably full range of words, it is impossible to ask for assistance; if you cannot remember, you cannot find your way home; if you cannot interact socially, you cannot have a normal social life. Ability to cope is not easy to quantify but it is a real measure of the many facets of the brain's functions.

The American Psychiatric Association has tried to classify mental retardation using these two measures to define four levels of retardation: mild, moderate, severe, and profound. Mildly retarded individuals can live independently and have a job; moderately retarded can care for themselves but normally will require some help for this and can have simple conversations; severely retarded people need much help and will learn some skills in looking after themselves but, although they understand reasonably well, they will have difficulty in speaking; finally, profoundly retarded individuals may understand some simple speech but will require complete and continuous care.

These four levels seem reasonable measures of increasingly severe problems and there is a parallel scale of impairment of intelligence. The problem here is the difficulty of defining a measure of intelligence and the American Psychiatric Association has relied on standardized intelligence tests that give a numerical value—the widely used IQ test. I do not wish to enter, at this point, the scientific arguments surrounding the IQ test (these come later) but here let's simply assume there is agreement that the average IQ is 100. A mildly retarded individual will have an IQ of about 50–70: for the other three categories the IQs are 35–50 (moderate), 20–35 (severe), and below 20 (profound).

This seems a very reasonable classification. Obviously there are real difficulties in deciding the limits (for example, is a person a mild case of severe retardation or a severe case of moderate retardation?), but these are really rather unimportant because what is being described is a spectrum of effects from normal to profound. The important point is that all of us would recognize a severely retarded individual even if we could not say exactly where they were on the scale. There is no disagreement that some individuals are mentally retarded and so we do not have to be so concerned with arguments over the specific measures that have been used to define such cases.

Selecting a degree of retardation that is unarguable is a way of overcoming the problem of 'measure' of the mind that I mentioned earlier, and this means we can look at the possible effects of genes on mental retardation in much the same way as we can look at the effects of genes on any objectively measurable physical feature of humans. In the examples that follow, I have deliberately chosen conditions for which there is no argument about either the gene's involvement or the consequences to the individual. Then I go on to discuss how even these clear-cut cases show how difficult the study of genes and behaviour can be.

Too many genes are bad for the mind

In 1866, the English physician J. H. L. Down published a short one-page paper in a scientific journal with the title 'Observations on an ethnic classification of idiots'. Papers are how all scientists tell each other what they have discovered: we write down our results and discuss why, how, and what we have discovered or realized. Down had observed that a fairly small proportion of children were mentally retarded (his description was that they were idiots, a term we no longer use) and they had eyes that were rather reminiscent of those of people of Far Eastern origin; their eyes were hooded. For this reason Down believed they could be classified as being of Mongol origin (from a region of central Asia) and the condition, until relatively recently, was often called 'Mongolism'.

Down was totally wrong about a Mongol origin. From our modern perspective, he was wrong for a good reason—he did not have the tools for studying DNA that we have now; and for a bad reason—there was a prevailing Victorian belief in the innate intelligence and superiority of the white 'races' over all non-white 'races'. We now know that this condition, which has subsequently become known as 'Down's syndrome', is caused by having too many genes. How can somebody have too many genes? The whole point of cell division is to make exactly one new copy of DNA and pass just this one copy to the new cell.

This means that every cell should have exactly the same number of genes; most of the time this is exactly what happens. In life, though, just as in every other sphere of existence, mistakes happen and in Down's syndrome the mistake is rather large in terms of DNA and genes.

The 3000 million rungs of DNA in human cells is not one continuous strand and is instead broken up into 23 specific fragments, which are neatly wrapped in proteins within the cells. The protein and DNA packages were first observed many decades before we knew anything about genes. They were given the name 'chromosome' because they appeared as a coloured ('chromo') body ('soma') after the cells were appropriately stained and looked at under a microscope. Because each DNA fragment is packaged up into a separate chromosome, each DNA fragment has itself, confusingly, become known as a chromosome, even though it is not the coloured body seen under a microscope.

Each human cell actually has two of each chromosome—one passed to us by our mother, the other by our father—so there are 46 in total. In contrast, a child born with Down's syndrome has 47 chromosomes: one chromosome, about 50 million rungs long, seems to be present not twice but instead three times. Each chromosome pair is given a number from 1 onwards and it is one of the smallest, number 21, that is present three times in the Down's syndrome child. Because each different chromosome has a different set of genes, the Down's child must have an additional set of the genes that are normally contained on chromosome 21. We know there must be about 200–300 such genes because the entire rung order of the chromosome has been established—one of the earliest successes of the Human Genome Project.

Does having an extra copy of a set of genes matter? Normally, the amount of a protein that is produced in a cell is controlled very carefully because, often, too much or too little can have disastrous results for the cell. So having an extra copy of even one gene can produce too much protein, which in turn can alter the correct functioning of the cell; having three copies of all of the genes on chromosome 21 must somehow lead to changes to the cells of the developing embryo, which in turn leads to the baby being born with Down's syndrome.

What causes the mistake that results in three copies of chromosome 21? It happens by a failure in the control of the complicated dance of the DNA molecules during the cell division that makes the egg (in most cases) or the sperm (in about 5 per cent of cases). Normally all of the 46 chromosomes line up along the central line of the cell as it is about to divide. One set of 23 chromosomes moves to one end and the other set to the other end of this axis; the cell then neatly divides down the middle, leaving 23 chromosomes in each cell. What seems to happen in the case of Down's syndrome is that this process fails and two copies of chromosome 21 end up in one cell and none in the other. Surprisingly failures such as these are not as rare as one might imagine and individuals are born who have three of chromosome 18 or extra copies of the X and Y chromosome (the X and Y are special chromosomes because males have an X and a Y chromosome and females have two Xs). Having three of all the other chromosomes produces such disruption that the cell itself dies and so is never found in living humans: it is only by unfortunate chance that the extra genes on chromosome 21 are compatible with life.

Down concentrated on the mental aspects of Down's syndrome, one of the most striking characteristics of a Down's child. The average IQ of many Down's children is 55 but one child in 10 has an IQ score that is not far off the 'normal' average of 100. Given this range of abilities in Down's children, it is not easy to talk about a single set of 'symptoms' and those with the condition fall within a range from minor to severe. 'Average' Down's children often have difficulty in talking and by adolescence they may have the speech abilities of a normal 3 year old; many will die young from heart defects and blood cancers are also very common. Most remarkably, if the children survive into adulthood, virtually all will suffer from Alzheimer's disease (a very common and distressing condition with symptoms of loss of memory followed by disintegration of personality and death). What is particularly striking about this is that a gene difference on chromosome 21 is important in contributing to Alzheimer's disease (I discuss this in more detail in Chapter 10) and we can only assume that having three copies of the gene instead of two makes a Down's individual particularly susceptible to this unpleasant condition.

Geneticists do not yet know which of the 200–300 genes causes the mental problems in Down's syndrome. There are rare cases where only parts of chromosome 21 are present three times and these are useful in narrowing down the region, and therefore the genes, causing the mental retardation. This sounds like a reasonable approach, but as yet it has not been particularly successful because of the great range of IQ in Down's children, which makes it difficult to make a precise analysis of the relationship of the genes lost against the symptoms. Why should IQ be so variable? There are several possibilities but the most simple is that the children have different sets of gene differences located on chromosomes other than 21 (because they have different parents) and these can magnify or reduce the effects of the extra chromosome. This is a classic problem with human beings: how do you allow for the differences in our genes?

Down's children are common and I imagine that everybody has seen or known at least one if not several. Scientists have known for a long time—since the mid-1930s—that older mothers are most at risk for having a child with Down's syndrome; for mothers who are 30 years old, 1 in 1000 babies will have Down's but for mothers who are aged 40, 9 out of 1000 babies will be born with the condition. This is why such mothers are now offered (or should be, because this is not the case in many poor countries) a test while they are in the early stages of pregnancy. The test ultimately involves taking a few cells from the fluid that surrounds the baby to see if the extra chromosome 21 is present. The difficult question that arises out of these tests is one that is going to recur over and over again in my discussion of genes and how they affect human beings; it is simple and obvious: what will my child be like? Will he or she be badly damaged? We know that there is a great range of symptoms in Down's children: will my child be high or low IQ? Will my child have a normal heart? Will he or she get blood cancer or die of Alzheimer's disease at aged 47? Above all, will my baby learn to talk a little or a lot?

All the test will tell you is that these could be the fates of your baby but it cannot say which one is the particular fate of your child. The reason for this is that your baby is a unique product of unique parents and at present there is no way of quantifying the effects of uniqueness because it is the result of differences within many genes: this is the major problem with understanding the effects of gene differences on human beings and it is a difficulty to which I will return again and again. The lack of certainty of outcome is the major reason why many people will still choose to have a baby even in the certain knowledge it has Down's syndrome, because there is always the hope that the child will have a happy, loving, and content life, even if it is short.

Does Down's syndrome really tell us anything about mental retardation, or IQ and genes? Yes, it does, despite all of the problems I have discussed: it tells us with certainty that genes can influence the most sophisticated workings of the brain—our intelligence and our social abilities. There is no disagreement over this; the impairment to intelligence is real and significant. What it does not yet allow us to understand is how the genes do this nor which genes are important.

Down's syndrome is unique in being caused by an extra copy of a whole chromosome 21. We also know that there are many parts of other chromosomes that occur in the wrong number of copies in the cell and can also cause mild mental retardation of very unpredictable severity. These extra regions of DNA are rare—perhaps 1 in 100 000 babies—but involve many different regions; part of chromosome 11 in one case, part of chromosome 9 in another. This reinforces the point that extra genes can influence intelligence but again does not tell us which of the many genes in DNA are involved. A clue to under-standing the effects of specific genes on intelligence comes from looking at the DNA in children who are born with fewer genes than normal—the topic of the next section.

Too few genes is just as bad: eyes without colour and kidney cancers

There is no obvious connection between your kidneys and the coloured part of your eye—the iris—but about 1 in 10 000 babies are born mentally retarded, with a peculiar type of kidney cancer, and with all or part of the iris missing: a syndrome given the acronym WAGR. Why this curious jumble of symptoms occurs is proof that too little of some proteins results in mental retardation.

If having too many genes is bad, then having too few is likely to be equally unfortunate: this is indeed the case. One of the serious effects of chemicals and radiation on DNA is to cause breaks to the DNA ladder and pieces to be lost, resulting in the loss of some genes. Instead of two copies of the genes in the cell, therefore, there might be only one. A gene cannot know that it has to make twice as much protein (how could it, what would tell it?), so almost always losing a copy of a gene causes the cell to behave abnormally. This is exactly what happens to genes normally found in a specific region of human chromosome 11. We know the identity of several genes within this region of a few million rungs; one is called WT1 and a second PAX6—the same gene that encodes the protein PAX6 I mentioned in Chapter 2 with reference to making an eye. Both of these genes make gene regulators: WT1 is needed to control the development of a normal kidney (not making enough WT1 protein means that the kidney cells grow uncontrollably and become cancerous) and PAX6, as I discussed before, is required to make the iris of the eye. Loss of the two genes is why the babies have no iris and frequently develop kidney cancer. Why are the babies also mentally retarded? Because at least one other gene needed for normal intelligence has been lost from the DNA ladder.

At present, we do not know how many other genes are contained in the region that is lost in these children. My own research group is currently working on exactly this problem as a part of the Human Genome Project that I intro-duced in Chapter 2. We are working out the exact sequence of 14 million rungs, taken from a normal person, that corresponds to the region that is lost in these children (the 2 million or so rungs is the smallest region that is lost). Once we have done this we will be able to identify all of the genes and then work out which gene (or perhaps more than one) is important in causing the mental retardation. We think that our research is important not only because it will help understand this very distressing and serious condition but also because it will throw some light on exactly which genes are required to make a normally intelligent human being.

This last point brings out the major difficulty in what I have discussed so far. Both Down's children and the kidney/iris children have lost many genes. We know of many rather similar regions where losing DNA results in children with mental retardation, and always the regions are large. This raises a difficult problem: perhaps the brain is so complicated that single genes do not have any effect; perhaps it is only by gaining or losing many genes that an effect is seen? That this is not so, and that the effects of losing just a single gene can be startlingly specific, is shown by the case of children who have the remarkable William's syndrome.

Elfin children: William's syndrome

Children with William's syndrome are really quite unusual and challenge many of our beliefs in what we mean by mental retardation. Most children with the syndrome have IQs in the 50s—well into the mild/moderately retarded state, but they also seem to delight in story telling and passing information, and are extraordinarily friendly. In fact, they are almost always more developed in their talking skills than normal children of the same age.

William's syndrome children commonly have an unusual face, frequently described as pixie-like and with a wide grin, and often have heart disorders. It is a very curious mix of loss and gain of abilities, but the most curious loss is in the ability to comprehend the relationship of objects to each other and to the space the objects occupy. If you were to ask a William's syndrome child to copy a pile of bricks put together in an organized shape, the child would be unable to do so; the ability to think about objects in relation to each other appears to have been lost.

What has happened to the DNA ladder to make this constellation of unusual abilities? The answer is that a small region of DNA has been lost. The actual amount is different in different children, but probably in every case two genes have been lost. One gene makes elastin, a protein that is an important part of the tissue that connects cells together. The reason William's syndrome children have heart problems is because the heart's connecting tissue is weak and prone to burst, causing a heart attack. Its loss also accounts for the elfin features because the shape of the face is in part formed by the way the tissue connects together. Elastin is also important to the development of proper nerve connections because the nerve cells actually move over the elastin surface and a loss of the right amount of elastin may cause miss-wiring of the nerves. This may partly explain why these children have such unusual minds.

The second gene that is always lost in William's syndrome children is perhaps more important than this. It encodes the protein called Lim1 kinase, or LimK for short. LimK modifies a gene regulator called Lim1 and hence controls its function. Lim1 is an important gene regulator involved in controlling the formation of the head (we know this because mice lacking Lim1 develop without a head). Lim1's activities are in part controlled by LimK and so, in the absence of LimK, Lim1 will be made in an aberrant fashion. Given the importance of Lim1 to the brain's development, it is unsurprising that there is a disruption of normal brain function in William's syndrome children: in fact, the front parts of the brain are slightly altered, the front-most part being slightly larger and the part behind this smaller than normal. Indeed, the slight enlargement of the front part may contribute to the overdeveloped language skills and perhaps the reduced parts contribute to the IQ impairment seen in the syndrome—but this is wild speculation and we cannot really be sure; the link seems just a little too simple to be true!

Speculation is a dangerous thing in genetics: the simple truth is that often we are wrong unless we have solid evidence to support what we are saying. Nevertheless, I suspect that William's syndrome children may be another key to understanding human intelligence. IQ tests are used as one measure of intelligence (there are many different IQ tests, of course, because there are many different ways a child can demonstrate his or her intelligence). One major part of many tests sets out to find out how good your mind is at relating objects. One question that tests this ability asks which object cannot be flipped, rotated, inverted, or warped to be the same as all the other objects. Of course, this is exactly what William's children cannot do and so it raises the following possibility: if William's children have less than normal amounts of LimK and develop so that they are less good at these sorts of intellectual tasks, is it possible that differences within the LimK gene may result in alterations to the brain that result in an enhanced ability to handle the relationship of objects? Perhaps one reason why people (and I want to stress this is only one reason) have a greater intelligence than others is because of this. To the best of my knowledge, no one has ever tried to test this by analysing LimK in very clever people. This experiment, as you will see, is much more difficult than this simple view might suggest (I'll discuss why in Chapter 8) but certainly worth thinking about.

As yet scientists have identified few genes whose loss results in these sorts of changes to the brain, but it is clear that many different regions of DNA must contain such genes because of what we know about the abilities of people who are born lacking regions of DNA (as in the case of the kidney tumour/iris/mental retardation I mentioned earlier). Indeed, the brain seems to be greatly altered by gene loss. This is perhaps unsurprising because the brain is the most complex organ in the body and so is probably the first to be affected by such losses; complicated structures seem to be the most sensitive to change. Support for such a view comes from evidence that several genes seem to be involved in fairly common cases of mental retardation and aberrant behaviour.

One gene at a time

Single-gene differences that can alter intelligence and behaviour are now quite well known—scientists have known about some for some 30 years or so. Mental retardation is usually caused by complicated events involving, amongst other causes, an unknown number of gene differences. One of the commonest specific causes, though, is a product of curious changes in a gene called FMR-1, which, although first discovered in 1991, still remains of unknown function. DNA differences in this gene are found in about 1 in 5000 babies and result not only in the failure of the gene to produce protein but also causes the DNA to break within the FMR-1 gene itself. Rather unusually this breakage seems to happen during development, giving rise to cells with and without the broken DNA. It seems to be the lack of FMR-1 protein that causes the mental problems because there is some evidence it is a gene regulator; we do not know for sure.

About 1 in 20 000 babies are born with a disorder called Lesch–Nyhan syndrome (named after the first doctors to describe its unpleasant symptoms), which is caused by differences in the gene that encodes a protein called HPRT—part of the chemical factory of a cell. HPRT converts one small chemical molecule into another; the details are unimportant but what is important to the cell is that the final product is part of the chemical changes that have to be made in a cell so that it can make the component parts of DNA. The HPRT gene either fails to make any HPRT protein or makes one that cannot carry out its normal role. The result of this seems to be that particular types of nerve cell do not develop properly. The impact on the baby born with this gene difference is devastating: they are mentally retarded, with moderate or severe learning difficulties and speech defects as well as extraordinary self-injuring behaviour. The children will bite their fingers or lips so badly that they will, quite literally, eat parts of themselves, even though it is very painful to do so. Why this should happen is not understood but presumably relates to the damage to the nerve cells.

The changes to the HPRT gene result in an untreatable condition. It is very different from another rather common disorder caused by gene differences involving a protein called PAH. Babies born with differences to the PAH gene can have IQs ranging from near normal to below 50. The normal role of the PAH protein in the cell is to help (along with other proteins) break down an amino acid called phenylalanine that occurs in protein taken in as food. In children born with no PAH, unusually high levels of phenylalanine accumulate in the body and become poisonous, causing damage to the brain. Because the condition results in phenylalanine in the urine, it is called phenylketonuria, or PKU. PKU is unusual for a disorder caused by gene differences in that it is moderately easy to prevent by ensuring that the newborn child follows a low-phenylalanine diet. For this reason every child born in the UK, amongst many other countries, is tested by taking a simple heel-prick blood test.

All of these examples show that genes can influence the brain and that differences to genes can cause mental retardation. There are perhaps 100 different conditions that include mental retardation in their symptoms and in most cases we don't know which genes are involved. The key question we must now face is exactly what is it we are observing when we see a child with mental retardation? Are we looking at some very specific effect of a protein on brain function or are we looking at a more general effect, perhaps disruption of normal development, that results in more general brain 'damage' and therefore mental retardation. This point will become important when I discuss the effects of gene differences on intelligence (Chapter 8), and so I'll pursue it a little further here.

The sledgehammer or the loose screw?

Mental retardation is obviously a sign that something has gone wrong in the workings of the mind. This could be due to a problem in brain development, such as seems to occur in William's syndrome children or the case of HPRT, or it could be due to poisoning of cells after reasonably normal development, which we see in PKU. In any event, all of the above examples tell us that genes can influence intelligence. Or do they? Is it possible that the explanation is a bit simpler than we might imagine? Consider an absurd argument; I'd like to suggest that a sledgehammer could control your intelligence. My argument is that if I were to smash your head with a sledgehammer I would be very likely to reduce your intelligence quite dramatically and abruptly. This is, of course, true but what is absurd is that the sledgehammer is not directly influencing intelligence, it is just reorganizing your brain in a very unpleasant and non-specific way. In contrast, perhaps there are gene differences that act rather like a loose screw in a machine: if it falls off, a very specific part of the machine fails; if it is tightened, perhaps the machine will work a little smoother. The key difference here is the word non-specific: there can be no doubt that a sledgehammer could control intelligence but only in the most non-specific of ways, in sharp contrast to the specific screw.

Does the evidence support the idea that genes control intelligence specifically, the screw analogy, rather than by changing the behaviour of cells in a general way, the sledgehammer analogy? This may seem an unnecessarily detailed question. After all, if a child is mentally retarded, why does the mechanism matter? Well, I can think of two good reasons: treatment and high intelligence. If we wish to treat a disease of any kind we must understand its cause because it is only by doing this that we will ever be able to design the appropriate drugs or other therapies (I'll discuss these in Chapter 8).

The second point relates to a deliberate bias that I have introduced in my examples so far. Until now, I have deliberately concentrated on gene differences that can cause mental retardation, but if these gene differences can be shown to cause specific modification of cells and regions of the brain, rather than non-specific effects, then we open a second possibility, which I briefly discussed earlier. Many proteins must be involved in forming the cells, regions, and functions of the brain; in the case of William's syndrome, loss of LimK results in changes to quite specific abilities. Is it possible that differences in some of the proteins involved in this region of the brain, other than LimK, may result in increases in intelligence rather than reduction?

This idea could have an important bearing on views of intelligence: at one extreme, we could argue that it is only the sledgehammer approach that can reduce mental abilities and there are no specific effects and so no prospect of increasing intelligence.

I do not think the evidence is very strong so far in either direction: the examples of PKU, HPRT, and the FMR-1 have the features of rather general effects. HPRT and PAH are indeed made in virtually all cells of the body and FMR-1 is found widely in both brain and cells that are not part of the nervous system at all. This makes it hard to see why the effect of the different proteins could be specific without suggesting some indirect arguments, the best of which is that perhaps what we are looking at are changes that do indeed affect every cell but that these influence specific parts of the brain most of all. I suspect that this may turn out to be exactly what is happening and perhaps the best way of looking at the effects of gene differences is as a hybrid of the sledgehammer and the screw: perhaps it is like a very sharp and pointed ice pick, selectively changing a small region in a non-specific fashion!

This really does not get us much further. There is still no good evidence for specificity and so we must seek another way to explore the possibility of gene differences causing specific changes to how the brain works. I think the problem we face at the moment is that mental retardation is much too broad a measure of the brain's function. There may be many ways, specific and non-specific, that a brain can be altered to cause mental retardation, but what if we could show that changes in very specific parts of human behaviour are associated with differences in single genes? We have started down this route already with William's syndrome where the overdeveloped language skills are a specific component of the brain's functions. What I want to do now is to look at the very few cases where such changes have occurred because they will show that some gene differences can be quite specific in their effect and this will also lead us to the much more intriguing, and perhaps frightening, idea that our normal behaviour is influenced by genes.

The killer-rapist mouse and other stories of violence

Violent people frighten us: this is not a profound statement. We are surrounded by many signs that violence and human beings are never that far apart, and huge numbers of words, thoughts, pages, and actions have been directed at understanding and explaining why we humans have this extraordinary capacity for destruction. It is far from certain that we are in any way unique in this as a species but this doesn't really matter; all that is important is to answer the question as to why some people are, and some are not, violent.

I don't think that great amounts of speculation are particularly useful on this topic; everyone seems to have their own ideas and theories, and nothing seems to progress much further. What I want to do is explore the possibility that some forms of violent behaviour may result from gene differences. Here I want to touch on the good evidence that a specific set of violent behaviours might be strongly influenced by genes; the more general problem of how this fits into our view of humans and human society I will leave to later (Chapters 10 and 11). If we can show that genes influence specific behaviours, then we are going part of the way to deciding that genes can influence the brain in a specific rather than non-specific way.

First, and briefly, I wish to mention a rather remarkable finding that was seen when genetic scientists started to study a particular chemical in the brain of mice. The chemical is very simple and is called NO (for nitric oxide). NO is made by a protein called NOS (which stands for nitric oxide synthase) and is used by many cells to communicate with nearby cells. As I hope Chapter 2 made clear, studying this sort of communication is important in understanding how the body is organized, and so making a mouse that no longer had a NOS gene was an obvious experiment; the geneticists used transgenesis experiments to do this (I discuss this technical approach in Chapter 9). Such a mouse was made with no NOS genes at all and the results were really quite staggering.

The mice seemed relatively unaffected—they looked normal—and this must have been a bit of a disappointment to the scientists. After all, if NOS is so important, how can mice look normal without it? Then something very unusual was seen: dead mice that had obviously been killed by their fellow mice began accumulating in the cages. So the scientists started to watch the mice very carefully and discovered two rather unpleasant things. The male mice were very aggressive and would fight, to the death, with their neighbours, which is most unusual in mice. Normally, they will fight but not to the death; the winner will rule the cage while the loser cowers in the corner submissively. These NOS-deficient mice killed.

The second strange behaviour was over sex. Mice mate generally late at night, and, having mated, the males will leave the females alone. Not so the male NOS-deficient mice: they would mate and repeatedly attempt to continue this even when the female tried to reject them. In short, these were killer and rapist mice.

Two features of this experiment are very important. First, because of the way the experiment was done, the NOS-deficient mice were in every other respect genetically identical to the mice that still had the NOS gene: all of the other genes were identical to the rung. Here we have a near-perfect experiment—two identical mice with only one difference; the presence or absence of the NOS gene. This means that the complicating effects of differences within any other genes in the animals could be eliminated because they were identical in both types of mice. This is completely unlike the situation in humans, where any two of us must have many millions of rung differences, which means it is very hard to tell if it is a difference within a specific single gene, or the combination of differences within the rest of the genes, that causes the effect. It follows that this mice study is a much clearer experiment than could ever have been achieved by observing violent humans.

The second, and critical, point is that these NOS-deficient mice were brought up identically to their non-NOS-deficient brothers. We know that it is possible to make a human violent by bringing him or her up in violent circumstances.

There have been dozens and dozens of studies that show that if a child is brought up in an environment where violence is both normal and acceptable, then it is more likely that the child will itself be violent. In contrast, mice live in laboratory cages in carefully controlled conditions and are brought up as close identically as any two individuals can ever be—much more similarly than humans could be. Even with this constant environment, the NOS-deficient mice were violent and their non-deficient brothers were not, making it virtually certain that it is the gene difference that causes this behaviour and not upbringing and environment.

We do not really understand why NOS deficiency causes such behaviour: NOS protein is present in many different cell types and so we, again, have the problem of specificity. Is it that the parts of the brain, wherever they may be, that control violent behaviour are particularly sensitive to lack of NOS? We do not know. There is a great deal of work being done on these mice to try and understand how and which regions of the brain may be affected. If we can establish this, then perhaps we can understand what is happening in some humans. At present, we have no evidence that there is a difference within the NOS gene of any human that might contribute to his or her violent behaviour, but geneticists are looking. If they were to find such a difference, then the social difficulties that are opened are substantial (these I discuss in Chapters 10 and 11). For now, we need only draw one conclusion: some individual genes can influence specific behaviours in mice, and NOS is just one example. Of course, aggression is a complicated behaviour and so, unsurprisingly, these single gene differences change only part of the aggressive behaviour. Wild mice are somewhat more aggressive than most laboratory mice and this supports the idea than many different genes can influence aggression.

This is all fine but men, with few exceptions, are not mice. This is perhaps not a profound observation but unfortunately it is rather likely that as it is for men, so it is for mice. We now that this is true because humans can lack a protein called monoamine oxidase, or MAO. MAO is involved in making a chemical that is important in signalling within the brain. There are actually two forms of MAO in humans, A and B, and a few people have been identified who lack the MAO A form. Understanding how the lack of MAO has affected humans is a story that is still unfolding and is, I suspect, the forerunner of many similar stories that will follow. The individuals who lacked MAO were all related—members of one large family with the striking feature that many of the males were unusually aggressive and violent, as well as being prone to outbreaks of arson, attempted rape, and exhibitionism. There seemed to be a simple relationship between this difficult behaviour and MAO: the men who were violent lacked MAO.

When these studies were published in a scientific journal in 1993, they caused an immediate and widespread reaction because a simple interpretation was that violent men might all lack MAO and therefore criminality of this type could be caused by the gene difference, which would have very important implications for our society. It was unsurprising that the study came under immediate analysis by psychologists: many questions were raised—that the characteristics of the men might have been badly described, for many different reasons, and that the relationship of the lack of MAO to the behaviour was therefore just a chance occurrence. These sorts of argument are common in science, as different scientists test out the strength of evidence for or against a particular explanation. The observation might well have languished, lost in this mire of conflicting opinion, had not a second group of geneticists made a mouse without MAO: sure enough, the mice were violent. This suggests that the observation is correct: MAO loss seems to change some specific function of the human and mouse brain that normally controls impulsive, violent behaviour.

From the evidence I have presented, you may think that criminal violence in any human being can be explained by lack of MAO and that criminals lack MAO (and, indeed, the original discovery prompted headlines that suggested this possibility). You would be wrong to do so for reasons I'll discuss later. For the moment I want simply to argue that the examples of NOS and MAO add weight to the idea that the brain can be specifically influenced by genes. I'd like now to discuss the evidence that genes can affect behaviour regarded as 'normal' in some people, not just behaviour that we think is abnormal or unusual. The area we are going to explore is personal behaviour and why we make some of the choices we do.

Genes and personality

7

I deliberately created a future history for Jean Battler and Jeanne Dream that does not tell us much about their personalities because I wanted to concentrate on other aspects of their lives but, of course, each would have had a distinct personality because possession of a personality is an inextricable feature of being a human—it would be as impossible to conceive of a human without a personality as it would be to think of two humans with absolutely identical personalities. The question we will focus on in this chapter is very similar to the questions we have been asking about the physical differences between each of us: how much of the difference is due to gene changes and how much due to environment?

Who was the first person to try bungee jumping? Who first leapt off a building with a parachute and invented the sport of BASE jumping? Whose idea of fun is to climb unprotected up a 150-foot cliff, knowing a fall was not only likely but almost certain to be fatal? Who throws themselves off an equally high waterfall into a bowel of rapids? Crazies, that's who; everybody who does these things is crazy—they have a death wish at the very least, and are totally mad to boot.

Well, many years ago I was a serious climber who loved climbing in winter: we would climb on frozen waterfalls and in icy gullies and I think many people who saw us were equally convinced we were mad, but I know that I was not. I used to explain to people who wanted to know that climbing was one of the few serious games we could play. If you lost at climbing, you did not just lose the game, you could, at the limit, lose your life. To me this was a huge attraction: not that I wanted to lose my life—far from it—I simply felt an overwhelming sense of achievement and satisfaction that I could use all my skills and stay alive. This is why many people have hobbies that are dangerous. To most people they are a source of genuine puzzlement: why do they enjoy this danger? Why don't they enjoy safe things?

This sort of question is not hard to answer at a superficial level; the reason some people like dangerous sports is because they have a personality that leads them in that direction. We all know the type: the person who loves driving cars at 120 miles an hour, loves skiing flat out down the mountain, or dashing head first into any new, exciting situation. We may shake our heads—there goes mad John again—but we know that this is how John is and we accept this because it is John's personality. We know that John may be wild and adventurous and so is Jane, but we equally know that Jim and Mary are quite the reverse—never happy in unusual situations but wonderfully welcoming at home and marvellous company, because these are their personalities. And, of course, there's that interminably boring friend of Celia's—he's so dull I could die of boredom just thinking about him, because that's his personality.

Just as all human beings look different, all human beings have different personalities. You do not have to carry out a particularly sophisticated survey to prove this; it is wildly obvious from your own experience. We also often believe that personality tends to colour a person's whole approach to life, even though a slightly more scientific approach may be necessary to support this view. We accept that a very conservative personality, for example, may often be conservative in a whole set of areas—clothes, food, films, spouse, home, politics, and so on. Of course we know that there can be exceptions to this rule and, if we are wise, we rarely should act on generalizations or expectations of behaviour to specific circumstances. Nevertheless, there are those who climb mountains and those who are equally happy reading about it in the Sunday newspapers, tucked up in bed with a cup of coffee. Who is foolish enough to claim who is happier?

What is the scientific evidence for the view that our personality colours all of the facets of our lives? If true, then what is the source of our personality? Where do we get it from—are we born with it or is it the product of our experience? The accumulating evidence is that a personality is something that in part we are born with and in part we acquire, and the part we are born with is the product of our gene differences; I'll discuss the evidence for this next.

The cultured, agreeable, conscientious, neurotic extravert

The first obvious difficulty to working on personality in humans is how do we measure it? How do we distinguish the mountain climber from the Sunday-morning coffee drinker? One answer is slightly surprising—ask them! Almost all research in this area is carried out by asking people to agree or disagree with statements such as 'I am never shy when I meet new people', or 'I am always shy when I meet new people', or 'I am easily angered'. You can see how, by answering a few dozen or a hundred such questions, it should be possible to test if there is an overall tendency in the answers from a single person. For example, mad John whom I introduced above, who loves leaping in with both feet into any situation, is quite likely to thrive on novelty and will tend to answer accordingly.

Out of this mass of information, generated from the results of many such questionnaires, some general concepts have emerged that are broad descriptions of personality. One current measure has personality as five 'dimensions' or components: extraversion, neuroticism, agreeableness, conscientiousness, and culture (this is openness to experience). Each of these five dimensions can be subdivided to include more detailed characteristics: for example, extraversion includes liveliness, sociability, and impulsiveness, whereas neuroticism includes moodiness, irritability, and anxiousness. The questionnaires are designed to try and identify the relative weight of each of these in any individual's personality. This seems a slightly dubious prospect but one major reason why the questionnaires are believable is that they are very reproducible even over a space of 20 years or so. Of course, there are other measures of personality: a second one uses four dimensions to define personality type: novelty seeking, harm avoidance, reward dependence, and persistence.

These sorts of questionnaire can be much misunderstood because they seem to be so rigid in their answers. Many people refuse to believe that they have a fixed personality by arguing that they can choose to do anything they want at any time and therefore personality measures are an illusion. Of course, having a particular personality type does not fix you in any one behaviour pattern; how many times have we heard someone say 'it was so out of character' of a person's actions, and this is a great beauty of knowing people well—we can appreciate them both in character and out of character. The important point is that everybody has a tendency to, and not a certainty of, acting in character and it is the tendency that personality questionnaires try and measure.

Most research has focused on the effect of genes on extraversion and neuroticism: large studies on twins and adoption have shown that about 40–50 per cent of the differences between individuals is due to the effect of gene differences, with extraversion being slightly more strongly influenced than neuroticism. More detailed analysis suggests that some of the subtraits of these broad characteristics may be more influenced by genes than others. Differences in thrill-seeking are about 60 per cent due to gene differences in one twin study and about 50 per cent in a study of adopted identical twins.

There are, of course, many tenacious critics of this sort of research, primarily because people distrust the implications of what is found and the idea, quite erroneous, that the research suggests that genes rigidly determine personality. This latter point is clearly not true: genes influence personality but environment, in the sense of experience, has an equally important role. Nevertheless, one obvious criticism is that getting people to fill in questionnaires is a dangerous method of assessing anything; we must all have had fun by filling in a questionnaire twice, seeking to achieve opposite goals each time. To give a trivial example, magazines often have questionnaires seeking to find out if you are a good lover; often you are asked to select one of three questions as representing your prowess, or otherwise, and you do not have to be endowed with superhuman intelligence to see that asking a question such as 'Does your partner achieve orgasm (a) always, (b) sometimes, or (c) never' is capable of being answered with a degree of wishful thinking. This is indeed the classic problem with questionnaires: bias is very easy to introduce and very difficult to avoid. To overcome this, similar sorts of twin studies have been carried out with the personalities being assessed by observers, frequently the parents. The results from identical twins again show about 40–50 per cent of the differences are attributable to the effects of gene differences.

What can we conclude from all of this research? Quite simply that the differences between us are due partly to genes and partly to our life experience. The different parts of our personality may well be differently influenced but about 30–50 per cent of the differences are possibly attributable to gene differences and the remaining 50–70 per cent to our environment. Quite surprisingly the research also suggests that the physical environment shared between brothers and sisters—the shared environment—is not important but that the non-shared environment, the experience of each child individually, is more important in forming personality. How can we make this last comment? There are several ways but the obvious one is adoption studies. Adopted children have a shared environment and if it is this that contributes to developing a particular personality type, then they should be more similar than children brought up in separate homes: adopted children in the same home are not more similar than expected, which suggests it must be non-shared events that contribute to the personality. I personally find this result quite surprising—I'd have guessed that shared environment would be as important.

The science of personality is, like much of modern biology, relatively young. There is a large body of evidence that supports the views that I have taken here, but I do not think we have yet arrived at a definitive answer. I have not tried to cover substantial areas of this topic: for example, the relationship of personality to life experience is particularly fascinating. What if I was to suggest that genes influence your surroundings? Totally absurd would be the snap response, but think again. Surroundings in terms of genes are your environment and environment has two features. It can be inflicted on you: the 'I didn't ask to be born next to a nuclear waste tip' type of event or you can inflict it on yourself, as in 'I think I will go and flop in the sun and drink hugely too much wine'. What is unreasonable about suggesting that personality may well dictate choices in the self-inflicted category? I think this is very reasonable; indeed, one would be fairly surprised to discover that this was not the case and the consequence could (I stress the 'could' simply because the studies are far from complete) be that genes indirectly influence the sort of environment you choose to occupy via their influence on your personality. I think this is a quite fascinating thought simply because it turns everything on its head!

I believe that the balance of evidence strongly supports the idea that genes influence personality to a degree. The question is which genes? So far the evidence has been based on the most general of measures for genes' influence and we cannot tell if it is a few genes or many thousands (it is virtually certain it is not one). Is there any indication that just a few genes influence our minds? One way of establishing this would be to use linkage to show where a gene might reside or, more directly still, we could detect differences in the DNA of genes. I shall look, therefore, at two cases—that of the emotional mouse and the thrill-seeker—to get the first glimpse of the nature of the genes that might be involved.

The case of the emotional mouse

The first time I came across the 'emotional mouse' I have to confess that I burst out laughing; the idea of a little mouse weeping uncontrollably over the poetry of Shelley or the music of Tchaikovsky I found so wonderfully incongruous that laughter was the only appropriate response. Needless to say, I soon discovered that the case of the emotional mouse is a little less anthropomorphic and possibly one of the more important recent developments in behaviour research because it starts to use animals to seek genes that influence behaviour and may lead to the discovery of how the same genes might influence humans.

What on earth is an emotional mouse? When mice in a laboratory test are presented with a large, well-lit open space—called an open field—they respond in several different ways: some will cower in the corner, clearly terrified, and literally wetting themselves and defecating in fear, whereas others simply relish exploring the extent of field. These behaviours have been given the name of emotionality, even though it does not correspond precisely to our normal use of the word; mice with high emotionality are those that appear to be most fearful. In some remarkable experiments carried out over the years, mice were bred, in just 30 mouse generations, that were uniformly fearless explorers; other mice were, equally uniformly, fearful and terrified in the open-field test.

These experiments were actually quite easy to do; all that was required was to test a group of male and female mice for fearfulness and to breed the most frightened pair of mice together. Their children were similarly tested and again the most fearful pair was selected and bred; this process was repeated for 30 generations, after which all of the mice were uniformly fearful. Simultaneously the most fearless mice were detected and bred together each generation, giving rise to uniformly fearless mice.

What is happening here? The initial idea that underpins the experiment is that emotionality is a behaviour that is influenced by many genes—let's say 100 but, of course, we don't know. We can call these the 100 behaviour genes for the moment. Mice are no different from other animals and all mice will have different gene differences that arise quite naturally, just as I discussed in Chapter 2. Some of these differences will be in the 100 behaviour genes, simply because these are genes just like any other and so capable of being different. Some of the gene differences may cause the mouse to be more frightened in the open-field test whereas others may have the opposite effect and make the mouse fearless.

At the start of the experiment the mice will have a mixture of differences contributing to both frightened and fearless behaviour and so will behave averagely in the test. If you breed two such mice, then their children will be given some of the differences from each parent and each baby mouse will be given a different, chance collection of the behaviour-gene differences. So, entirely by chance, some mice will be given more of the fearless differences than the others and some will be given more of the fearful differences; testing all of the mice will show which are fearful and which are fearless. If we then identify the two mice that are most frightened, then we know that by chance they must have lost some of the fearless differences their parents had, and this makes them more fearful.

Now when we select these two mice, we would guess that they will still have some fearless as well as some fearful gene differences, but if we breed them and go through the same process, their children have a chance of loosing the fearless differences; if not their children, their grandchildren or any one of their 30 generations of descendants. By the time we have bred all of these mice, the final fearful mouse will have only fearful gene differences and no fearless. In an exactly analogous fashion, the mice selected for fearlessness will have only the fearless gene differences.

This experiment proves that emotionality in the mouse is a behaviour that is under the control of gene differences and also that environment has nothing to do with the result because it is the same in all of the mice—they all live in cages with very rigidly controlled diets, day lengths, and so on. Of course, the experiment doesn't at all prove that upbringing cannot influence behaviour: this would be quite difficult to do in a mouse, but you could certainly try and train a fearful mouse to be fearless. It is entirely possible, indeed I suspect likely, that upbringing could mask the effects of genes but I am unsure if this experiment has ever been done directly.

That the environment of mice can influence test results was shown recently in a study that looked at the ability of different laboratories successfully to score the same mice for the same behaviours. The results were encouraging—the laboratories were reasonably similar—but also discouraging in that there were clear differences. How subtle the differences could be was shown by the observation that one researcher was partially allergic to mice and so always approached the mice wearing a breathing apparatus that looked a bit like a space helmet. Mice, unsurprisingly, found this disturbing and hence reacted differently to being handled by this researcher than to any others. The message from this research is quite clear—it is important that observations be reproducible, or replicated (in technical language), in independent laboratories.

Is there any evidence as to which genes are important in defining open-field behaviour? We can conclude that there must be many such genes because it takes 30 generations to breed the two mouse types; if there were only one or two genes, then this could be achieved in just a few generations. The results unfortunately do not allow us to be more precise. To overcome this, researchers can do another sort of experiment based on the realization that two different strains of mice exist, one of which is fearful and the other fearless.

What do we mean by a strain of mice? All mice belong to the same species—

Mus musculus is the scientific name for the mice commonly used in the laboratory—but mouse strains are a vital tool in genetic research. A strain of mice is the result of many generations of inbreeding—mating brother and sister mice together. The effects of this is to make the mice more and more genetically identical. After 70 generations the chances are almost 100 per cent that each member of every pair of genes will be identical—either identically different or identically normal; this means that the mice are essentially identical in the same way as human identical twins are identical. Unlike identical twins, which are always the same sex, inbred mice can be either males or females and, even more unusually, you can breed an inbred male and female mouse and its babies will be genetically identical to each other and to their mother or father. These type of mice are called 'inbred' strains and many of them have been created in laboratories around the world; all have different features because of the pattern of gene differences they contain.

It has been reasonably easy to detect behaviours that differ between different inbred strains as well as more obvious physical differences such as coat colour. Differences in behaviour are not so surprising given that all laboratory mice differ from ordinary house mice, even though they are the same animal. If you try catching a mouse in your house you will have a huge struggle; they are agile and will leap several feet to escape, whereas laboratory mice never leap like this. It was one of the very first behaviours the early breeders of mice wished to eliminate and they did it by selecting the worst jumpers for many generations until finally they had the non-jumping laboratory mouse.

Geneticists have known for a long time that two inbred strains of mice (I'll call them B and C for simplicity) differed in their behaviour in the open-field and other similar tests: C mice are much more active than B. This is because the C mice must have differences to both copies of some of the behaviour genes, whereas B mice probably have fearful differences in their genes. If you breed a mouse with a C father and B mother, you will have a mix of genes; the mice will be less fearful than the B mice but not as fearless as the C mice. If the various breeding experiments are carried out correctly (the details do not matter here), it is possible to get mice that have been passed gene differences from either the fearless Cs or the fearful Bs. These mice were tested for fearless or fearful behaviour and then the scientists carefully analysed the DNA of the mice of each type. They could use DNA analysis techniques to identify which DNA came from B and which from C. All of the fearless mice must have received their genes from C and so they looked to see if they could identify regions of C DNA that were shared by all fearless mice and regions of B DNA shared by all fearful mice. These regions should contain the gene difference that controls the behaviour. This is exactly what was done and three regions were found to be C DNA if the mouse was fearless and B DNA if the mouse was fearful: the search is now on to discover exactly which genes in the three regions are responsible.

What the experiment has indicated, even before this final goal is achieved, is that there are probably far fewer than 100 gene differences involved in conditioning fearful or fearless behaviour because if the number had been this large then these sorts of DNA analyses would not work: when too many genes are involved, the ability to detect the C DNA in a fearless mouse is much reduced because each gene difference contributes such a small amount to the overall behaviour. The experiment implies that at least three gene differences are having a substantial effect even if there may be more, as yet undetectable in this analysis.

The interesting possibility arising out of this research is that there is a relationship of mouse emotionality to human behaviours. What if the same genes are involved in mice and humans, even if the final influences on complex behaviours might be very different? This possibility is not completely absurd and researchers working on the gene differences contributing to neuroticism in humans are now concentrating their work on the region of human DNA that contains the same genes as the mouse regions known to influence emotionality. We don't yet know the results of this research—finding the actual genes is very time-consuming. If genes contribute to controlling very complicated behaviour in mice, why shouldn't it also be true for human beings? It is to the evidence for this that I now turn.

Brains, chemicals, and behaviour

The human brain uses chemicals to regulate its activities. These chemicals are released by the ends of nerves and will trigger an adjacent nerve to fire off a signal; in principle, a very simple signalling system but one that has, over the millions of years that brains have been evolving, grown to very great complexity indeed. A family of these signalling chemicals includes the monoamine transmitters that are acted upon by the protein MAO, which I mentioned earlier. The chemicals seems to be involved in regulating how the brain responds in a global way to reward and pleasure, to learning, sleeping, and attention span. Because of the importance of these responses, many mood- and perception-altering drugs that interfere with their function have been developed—Prozac most famously of the anti-depressant drugs, but also cocaine and LSD. The powerful effects of these drugs suggest that in the brain the monoamine transmitters are very important molecules indeed.

One of the members of the monoamine family is the chemical serotonin. The amount of serotonin in the brain has a critical effect on mood and behaviour; we know this because there are anti-depressant and anti-psychotic drugs that work by interfering with the various process in the brain that control serotonin levels. It is also known that high levels of serotonin can induce anxiety and panic in both humans and animals; this is why lowering serotonin levels with drugs is a potent treatment for depression and anxiety. Consequently a particular interest has been focused on the proteins that control the amounts of serotonin in our brains; one of the most important of these is called the 5-HT transporter or 5-HTT (5-HT is just the general chemical name for serotonin and related chemicals).

It was with a great deal of excitement that scientists discovered that there were two forms of the 5-HTT gene in people. The difference in the gene was not in the region coding for 5-HTT—all of us produce the same 5-HTT protein—it was in the switch DNA next to the gene. Careful work showed that the amount of protein produced by one form of the gene (the long or l-form) was about twice that produced by the short form (s). Because of the way 5-HTT works, people with s-gene differences should have more serotonin outside of their brain cells, and perhaps this in turn would contribute to greater anxiety.

This is a simple idea that can be tested by studying people who score highly on the neuroticism scale of personality: they should have the s-form of the gene; people low in neuroticism should have the l-form. In a group of 505 unrelated individuals and 459 siblings tested for neuroticism and the type of 5-HTT gene they had, the l-form was much more frequently associated with neuroticism than the s-form. This sounds pretty dramatic—you can imagine the newspaper headline 'gene for neurotic behaviour found'—but if you look carefully at what I've just written, you will see a strange phrase, 'more frequently associated'. Why this phrase? It's for the very good reason that the 5-HTT gene difference is not enough to make you neurotic by itself.

If we assume there are 10 or 15 gene differences that are required to push you high on the neurotic behaviour scale, then some people may have no difference in the 5-HTT gene but differences to all others; they consequently score very high. Other people may have the 5-HTT difference but not differences in any other of the genes, and score very low. The reality is that the differences to the 5-HTT gene probably account for only 4–10 per cent of the differences between us. This might sound a small difference but is nevertheless important; if any characteristic is defined by differences within 10 genes, then the contribution of any one gene may well be extremely small—10 per cent or less. This makes for a slightly less-exciting newspaper headline—'scientists discover gene difference that causes 1/10 of the neurotic behaviour in some individuals but has no observable effect in others'; hardly the stuff of journalistic legend! I'll come back to this point later.

The serotonin system is central to many different functions of the brain and researchers are now looking for an effect of the 5-HTT gene differences on complicated behaviour. There are early indications that there may be a relationship of gene difference to depression. One problem with such research, though, is that you need a large number of people to study. I have already mentioned that the number of genes potentially involved in influencing behaviour means that a given gene difference does not have to be found in all individuals; this in turn means that you need to design your study very carefully to eliminate the effects of chance. In this case the chance we have to avoid is that, by accident, a gene difference is found in affected people more often than non-affected people. The only simple way of doing this is to have a large study group, which may be difficult to find.

There is also a potentially huge trap in some of these experiments—one that is all too often fallen into by researchers working in this area. You will recall that I said that the scientists studying the 5-HTT gene examined 505 unrelated individuals and 459 siblings. Let's think about the reasoning behind this experiment a bit more carefully. Why both groups? In the 505 unrelated individuals what you expect is to have some people who score high on the neurotic scale and some who score low; you then look to see if the 5-HTT s-form is more common in the high group than in the low one. If you find it, you assume there is a relationship between the s-form and neuroticism. It could not be much simpler than this! How could this assumption possibly be wrong? Very easily, as it turns out. Let's suppose the 5-HTT difference has no effect whatsoever on behaviour and is just one of the many differences in human genes. When we compare the high and low neurotics we are comparing two different groups of people, and so what if the groups were physically different; at the most extreme, what if one was made up of African males and the other was all Chinese females? Could the difference between the two groups be the result of the differences between these two very distinct sets of individuals rather than anything at all to do with behaviour?

This is, of course, a crude exaggeration—scientists are very careful to avoid this sort of extreme bias in samples. This type of experiment is called a 'case/control study' because you are comparing the 'cases' (the high neurotics) with the 'controls' (the low neurotics). The scientists are making the assumption that any differences between the case and the controls are related only to the differ-ence in neuroticism and not to any other difference. Unfortunately the differ-ences between humans are at present enormously difficult to identify and count, which means that the case/control experiment is potentially flawed and can often turn out to be wrong—simply because the controls were different from the cases for more than just the feature under analysis.

Sib analysis is far better than case/control studies, because sibs, in sharing parents, must themselves share a common background. In practice, then, sib analysis is a far safer and better strategy to use for this type of gene research. Why, then, do so many researchers still use the case/control approach? Simply because identifying sibs is more difficult, which I believe is no excuse at all. The high level of concern and, to an extent, fear of this sort of gene research means that scientists have no excuse for doing potentially flawed research. Gradually, though, experiments on the influence of single genes on behaviour are being repeated with better-designed experiments. (Note that the work I discussed above on the 5-HTT gene differences is very thorough and carefully planned and is a model for other researchers.)

One of the most intriguing genes to come under intensive scrutiny recently produces a protein that is important in another of the monoamine trans-mitters—dopamine. Possibly people like mad John the manic bungee jumper, who I used to illustrate the differences in personality, have a difference within one of their dopamine-controlling genes.

Mad John the manic bungee jumper

The four-character personality scale that I discussed earlier uses 'novelty-seeking' as one of its four main personality characteristics, and the five-character scale contains sub-characters that define similar features. Questions such as 'I often do new things just for the thrill' are specifically designed to measure a tendency to seek out, and enjoy, new events and Mad John the manic bungee jumper is not just the invention of this book—I have known a few Mad Johns. Several different research groups have studied people who have high novelty-seeking personalities and they have concluded that differences in one of the dopamine-receptor protein genes is associated with these personality types. Dopamine is similar to serotonin and affects nerve cells by attaching to a specific protein on the cell surface—a signalling protein called the dopamine receptor D4 (D4 because there are several different signalling proteins) or DRD4 protein. Dopamine, much like serotonin, is an important chemical that can stimulate euphoria in humans and exploratory responses in animals; it was for precisely this reason that the researchers chose to study it for differences. The association is much like that of 5-HTT; many genes must contribute to novelty-seeking and the DRD4 difference can be responsible for only about 3 or 4 per cent of the variation between humans.

All this looks exciting, but a more careful examination starts to reveal holes in the arguments. One study used a case/control approach and must therefore be suspect; a second study used case/control as well as sib analysis—this should be more reliable. But at least two other large studies subsequently failed to detect a similar association between dopamine and novelty-seeking, even though they used a slightly different design of experiment again. None of the studies is necessarily 'wrong'; the different groups simply used different samples, different approaches, and experiments to reach different conclusions—there may, for example, be an association in some human populations but not in others. Unfortunately it is difficult to distinguish these two possibilities and this can be done only by careful repetition of the experiments in new groups of people.

The need for experiments to be repeated by other groups before they are really believable is very important in all branches of science; there is no implied criticism of the previous experimenter's results, just the need for an extra dimension of proof. This requirement is of even greater importance when the effect of genes on very personal behaviour starts to be the topic of investigation. To illustrate this, I am going to consider the question of whether genes contribute to making a person sexually attracted to members of the opposite sex; and are there such things as 'gay genes'?

A 'gay' gene?

Homosexuality, despite what many ill-informed moralists may believe about the late twentieth and early twenty-first centuries, has been a feature of human societies for many, many thousands of years. Moralists would also like to believe that homosexuality is a voluntary state—somehow a homosexual 'chooses' to be what he or she will be; whereas others believe that homosexuality is a disease and that homosexuals somehow have been infected with, or inherited, their sexual orientation. Queen Victoria, it is alleged, refused to believe that female homosexuality existed, and this was the reason that lesbians in the United Kingdom escaped many of the more drastic legal measures aimed against male homosexuals.

Few subjects except, perhaps, perceived racial differences, have generated quite so much bitterness, controversy, dissension, and misery as the disagreements over homosexuality and its underlying cause. Geneticists enter this arena with great reservations: their research is irrevocably tainted by a history of misuse and abuse of what has, or claimed to have, been discovered and what this discovery means.

To illustrate this let's consider why the outcome of genetic research into homosexuality can be so contentious. First, let's suppose that there is clear, scientifically incontrovertible evidence for a contribution of gene differences to homosexuality. The various factions of opinion are roughly divided into those who believe homosexuality is a moral evil, those who believe it is a disease, and those who believe it is a matter of personal choice. In all likelihood, each one of these groups will have a very different attitude to the discovery of a genetic component to homosexuality. The first group may see it as a vehicle for identifying and eliminating homosexuals. The second group may see it as a diagnostic tool for treating a disease, and the third group may consider it an irrelevance. But all three groups will certainly have subsidiary concerns about how the other groups of opinions will react, and this will certainly generate tensions and fears. Looked at from this viewpoint, it is clear that the identification of a gene that contributes to homosexuality will be greeted with very substantial misgivings by at least parts of our society.

The concern is not hypothetical—experimental evidence suggests such gene differences exist. The evidence is from DNA linkage experiments. Researchers have found that if one of a pair of twins is homosexual, then 50 per cent of the time the other will also be homosexual; this decreases to about 22 per cent of the time if the second twin is non-identical or an adopted brother. These results were then followed by studies with classic linkage analysis (of the sort I discussed in Chapter 2) of several families that contained a number of homosexuals.

There is no conceptual difference in using linkage analysis to locate a physical feature, a disease, or homosexuality: in each case exactly the same approach is used because all that is different is the condition under analysis. Using this approach, the researchers found that at least one gene difference on the X chromosome of human DNA seemed to contribute towards male homosexuality, but did not affect females. Needless to say, there was a tremendous amount of interest and excitement over this discovery, predictably split upon the lines I have already described. But lost in all of this was actual reality of the experiment: all that was achieved was a linkage to a region on the X chromosome that contributed to, but certainly did not define, sexual orientation. Subsequently these same researchers published a study that confirmed their original observations but a second, completely independent, group failed to see the association.

This again leaves us in a dilemma. We can hardly resort to a democratic voting approach to this question and say two against one means there is a gene difference, because this would be allowable only if we could first identify the flaw in the dissenting study. How could the conclusion be wrong? One simple way would be for the actual families not to be quite what they seemed. The definition of homosexual is important if you are going to identify homosexuals in a family, so let's try to define a homosexual. How about somebody who is actively living with and having sex with a man on a regular basis? This is no good; many homosexuals do not live with other men. So forget about the 'living with'; why not just having sex with other men. Also no good; because some homosexuals may have been sexually active in the past but are not at present. Does a single homosexual encounter mean the person is homosexual? What about adolescent experimentation with members of the same sex? Are these sufficient grounds to define a man as homosexual? What about repressed homosexuality? How do we define this? Does looking at a man's body and thinking 'that's a wonderful body' make a man a repressed homosexual? If so, most men must be in this category!

The difficulties are endless. Unless you can measure something accurately, then it is easy for false linkages to occur. For example you could scrutinize all cases where the particular X regions had been passed on to a non-homosexual, defined by some criterion, and then reclassify the individual as a repressed homosexual based on an alternative definition of homosexuality. Researchers are acutely conscious of this danger and have tried to use a rigorous and objective definition of homosexual, but there is always a real danger that errors will occur. A real physical measure is always easier to obtain objectively than a subjective one.

So the question is, in the absence of the final proof of a gene with a difference, is this research actually identifying a region that must contain such a gene (or genes) or is it a result obtained by a chance set of coincidences? My present bias is to sit very firmly on the fence (scientists have to be better at this activity even than politicians because it is the correct place to be when faced with insufficient experimental evidence). We need more experiments, but even if the results of these support the association, I am very unsure if we will ever be able to show exactly which gene is different because the X region is very large and contains many genes. It is, of course, possible that the results of the Human Genome Project, telling us about all of the genes located in the region combined with knowing the function of some of them, will enable us to make educated guesses as to which gene or genes may be involved.

So does this mean that scientists have discovered a 'gay gene' as many news-paper reports suggested? Of course not: as is so often the case, the scientific truth is far less exciting than the headline suggests. I think it is worth simply listing what was actually found in the original study compared with what the headlines imply was found.

- There is a gene for homosexuality, found only in such men. *No! No gene has been discovered, just a region on the X chromosome region that should contain a gene (or genes) with a difference that contributes to this sexual preference.*
- The gene difference makes a man homosexual. *No! It makes a small contribution to the reasons why some males are homosexual: background, upbringing, and environment all have even greater impact than the gene difference on the X chromosome.*
- The gene contributes to the behaviour of all male homosexuals. *No! The homosexuals studied in this research all came from carefully selected individuals with a clear family history of homosexuality. The study suggests that the gene difference on the X chromosome contributes to a man becoming homosexual in these families, but this may not be the case in other families where differences in completely different genes might be responsible. Until we know what the gene is and how it is different we cannot study this question. Once (and if) we know this, we can then see if the same difference occurs in many additional homosexuals and knowing this would allow us to conclude, or not, if the gene difference contributes to the behaviour of all male homosexuals.*

You may be slightly surprised at these last comments. It would seem reasonable that if a gene difference caused a condition in a family, it would also be responsible for causing the same condition whether it was in a family or not. Surprisingly there are a number of clear examples where this turns out not to be true; probably the best example is that of breast cancer. There are families where many of the women suffer from breast cancer before the age of 40, which is rather younger than in non-family (called 'sporadic') cases. The actual genes have been identified by linkage analysis and subsequent research has shown that there are differences in either of two genes responsible for the cancer, called Brca1 and 2. The discovery of the genes was followed by analysis of many women who had sporadic cases of breast cancer with absolutely no family history of the disease—the commonest type of sufferer. There is no sign of either gene being different in these people, an observation that proves that there must be several different ways of developing breast cancer: the rare family form is associated with differences to Brca1 or 2 and the common sporadic form is caused by an as yet unknown process. To return to the case of homosexuality, it is perfectly possible that identifying gene differences in a family might well tell us nothing about the gene differences in 'sporadic' homosexual individuals.

We have very little idea of the differences between homosexuals and hetero-sexuals other than sexual preference. Several researchers have claimed to have identified physical differences in the brains of homosexual men compared with heterosexuals but it is unclear if these claims are really correct. Small numbers of brains were examined after death and it remains possible that chance can explain the differences. If there really are such differences then one would expect many genes to be involved. I think that the question of the relationship of genes and sexual preference remains something of an enigma. Perhaps what we have discovered so far is the rough location of just one of the many genes that can influence sexual preference. But how many genes and how much effect? This question is the key to understanding how genes influence behaviour and is of enormous importance in helping us understand how society should view these effects.

How many genes make a feature?

We know that gene differences are important in influencing some sorts of behaviour, but how many genes are involved? Of course, we cannot tell: it is only by linkage analysis, or similar, that we can start to find out how many genes may be involved but here we meet the greatest problem of all: what if we find 200 genes that are different? What if each contributes only a two-hundreth part of the effect? Could we see each of these genes by linkage? No, probably we could not because we simply do not have enough families or sibs to be able to pinpoint such a small effect—we would need hundreds of thousands and we are lucky to have 20 000! What we can expect to see is gene differences that cause reasonable proportions, perhaps 1 or 2 per cent, of the variation in behaviour: we call these 'genes of major effect'.

What if there are no such genes of major effect? Some people have argued that for the brain to work well requires all the genes in the body to be working well: alter one gene and you alter the working of the body, which could influence how the brain behaves or develops. This would mean that twin studies would certainly support the idea that genes are important to the working of the brain but we would never be able to find out which genes are the most important. The hope amongst behaviour researchers is that we will be able to detect genes of major effect—perhaps as has already been done in the case of DRD4 and 5-HTT—but we have no way of knowing if this is really the case.

I am really unsure if we will ever learn which gene differences influence the brain's behaviour; a few have been identified and probably a few more will be. Until we have the results of the Human Genome Project and have a better idea of the gene differences with each of us, I remain slightly pessimistic about the ultimate goal of this sort of research. The good thing about science, of course, is my opinion is irrelevant because, in the end, the experiments will either work or they will not. I am certain that my opinion will be replaced by experimental fact within a decade—and I will be delighted to be proved wrong!

To summarize: I think that human behaviour is influenced by many genes and that a few of these will be important enough to be detectable, but these will be a small minority. So what does this mean to us? Am I forced to behave in a particular way by my genes; are you? What does a 'moderate' influence of genes mean? There is probably no question more misunderstood than the influence of the genes: I turn to it next.

The features of the genes' landscape

The great misunderstanding about genes is the idea that genes act alone: they do not. They act with other genes and they act in the context of your body, your lifestyle, and your history. This is especially the case with genes that affect behaviour—we know that the infant brain builds its circuitry partially in response to signals it receives when young. All of the studies of behaviour that I have mentioned show clearly that experience is more important than gene differences in virtually every case. A moderate influence of genes may be 20 per cent of the difference between us but this means that something else is influencing the remaining 80 per cent and that 'something else' is your history and behaviour. Customarily this is called your 'environment' but I think this word might mislead when used in the context of behaviour, because the environment has come to be associated too strongly with nature and the physical world. In the context of genes and the brain, environment includes absolutely all aspects of your life from conception to the present: physical, chemical, psychological, and emotional.

Let's think about what the effect of environment means. To make this easier, I'll take an example, that of schizophrenia. The history of research into schizophrenia is long and a good example of how thinking about the brain has changed over the years. In Western societies, about one in a hundred people are diagnosed as suffering from schizophrenia even though the condition is difficult to diagnose with certainty because it has a wide and variable range of symptoms. Patients will often suffer from delusions, hear voices talking to them, and tend to jump from subject to subject when talking and to make unusual connections. It is a very debilitating condition for the sufferer and normal social relations may become very difficult; perhaps half the beds in mental hospitals will be occupied by schizophrenics and probably 10 per cent of homeless people are thought to have symptoms. In all, this is a condition that causes terrible damage and suffering and costs society a tremendous amount in both direct care and indirect expense—probably even more than cancer.

Because of the human and financial cost, schizophrenia has been the subject of intense and continuing genetic research, but this has really achieved relatively little. The most important observations so far, which I touched on in Chapter 2, were made using twin studies, which showed that identical twins had about a 50 per cent chance of both having the disease and non-identical twins about a 20 per cent chance. This is strong evidence for a substantial effect of genes on the condition but pinpointing the actual regions by linkage analysis remains very difficult; at least eight regions are contenders, but in every case there is controversy over the findings. At this stage, I think it is a safe bet that schizophrenia is in part a condition caused by genes, but how many is still unknown and, for the reasons I have just discussed, may remain unknown for a long time.

Although the twin studies provide strong evidence that genes are important to the causes of schizophrenia, they also are a graphic illustration of the difficulties knowledge of genes brings. If we think about a pair of identical twins, one of whom has schizophrenia and the other does not, we can ask the question, will the currently unaffected twin develop schizophrenia? In spite of knowing that the twins have exactly the same DNA differences, this does not allow us to be certain about the fate of the unaffected twin. All we can say is the he or she has a 50 per cent chance of becoming schizophrenic. This is exactly like walking up to the twin and taking out a coin and saying 'I am going to toss this coin and if it lands heads (a 50 per cent chance) I will say you will develop schizophrenia, and if it lands tails you won't'.

No analysis of schizophrenia genes will be any more accurate than this because the two twins are genetically identical. In short, analysing gene differences cannot tell us if the individual twin will develop the disease, simply because there are two things needed to develop schizophrenia—gene differences and environment. All of the genetic analysis allows us to say that the twin has a much greater chance of getting schizophrenia that someone who has no twin with the disease—about a 50 per cent rather than a 1 per cent chance—but this is a poor piece of information for the unfortunate twin. Will I or will I not get this terrible disease is the question that the twin need answered and what does he or she get? The answer 'perhaps' and a toss of a coin!

This is the common problem with gene differences and complicated behaviour and disorders: genes never act alone, they always act with the environment. In practice, this means that geneticists can give only general not specific advice: you have a 50 per cent chance of becoming schizophrenic or a 10 per cent chance of becoming diabetic. This is not what people either want to hear, because of the lack of certainty, or expect to hear, because of the perceived power of gene-based diagnosis.

This is a key point to understand: for genes and behaviour, the effects have been far less than 50 per cent. This means, for example, that if you have a DRD4 gene difference it does not mean that you are a thrill-seeker. Instead it simply means that if I looked at many thrill-seekers I would detect a greater number of different DRD4 genes than in the same number of non-thrill-seekers. This in turn means that I can say absolutely nothing specific about your behaviour based on a knowledge of the DRD4 gene difference you might have; I can merely say that it is a little more likely that you will be a thrill-seeker than not, which is a rather meaningless comment, because, of course, hidden behind all of this genetic analysis is the other reason for being a thrill-seeker—environment. Perhaps you were the child of thrill-seeking parents and were brought up to be like Mad John; you knew no better and so it is natural to you.

So this is a very major lesson to learn about genes and human beings: genes can only occasionally tell you with certainty what you are and what you will become; more often than not, they tell you what you might be and what you might become. Understanding this is the key to unlocking the truth behind the headlines in newspapers and in inflated claims to have understood the basis of some human condition or disorder: we understand part but never the whole and so while we study genes alone we must remain ignorant of the ultimate outcome.

In conclusion

Everything I have discussed in this chapter drives home the point that genes are important in influencing our minds. In the early stages of life, genes craft our brains and our bodies and, quite literally, make us humans; there is no environmental effect on this most fundamental process except in the most extreme case of damage. We are born with a brain that is human and that will continue to develop as that of a human. We do not have any solid idea of exactly which genes achieve this remarkable goal, but we know the identity of some and can guess that many more must be involved. The strides that are being made in research into development are enormous and I think it is likely that within a few years we will know a great deal more about how the most complex parts of our brains develop and the way genes control this.

The influence of gene differences on our behaviour and personality is far less powerful; many people find the idea that genes influence behaviour at all to be disturbing and have even suggested that research into this is so socially sensitive that it should not be allowed to continue. This fear is based on a real failure to grasp the reality of what genes do to us; all of us are aware that all humans have personalities and that these personalities are different. Think about your friends and acquaintances for a moment: I would guess that they have many different approaches to life and its problems and excitements; perhaps I can even make some guesses as to what general types of personality they may have. The range will extend from those that relish new challenges to those that fear change; some will be gregarious and some more private; some will be emotional and sensitive and some more hidden; some will be prone to panic in a crisis and some will stay calm and controlled. You will expect each of the people you know to have the particular personality he or she has but you will also know that they can step out of this character on occasions. It is equally likely that you are aware of your own personality and know that you tend to respond to events in a particular way: you are, I hope, happy with this personality.

In a sense, then, our actions are determined by our personalities and this seems to be widely accepted. But why are we less happy with the idea that our personality is in turn determined by gene differences? Partly the unhappiness is based on an erroneous view that genes are acting with an iron fist and that somehow this implies we have no control over our behaviour: I hope that you will realize that this is not the case.

Perhaps another way of exploring this question is to ask why people are happy with the idea that genes have nothing to do with behaviour. What's so comforting about the idea that it is environment alone that has produced our behaviour? Is it because we feel that potentially we can change the environment and perhaps reverse the behaviour? What if the environment has already influenced the development of our brain during early life and this moulds subsequent behaviour; what if the early influence is irreversible? If it is, then we are faced with the possibility that environment is as inescapable as genetic inheritance. If we can research into both contributions to behaviour—gene differences and environment—then we may be able to tease out the gene differences that make some individuals more susceptible to the environmental causes of some behaviours. Paradoxically the greatest hope for gene research may be in the potential for environmental modification to influence outcomes.

The most general conclusion I can draw from the research and ideas I have summarized in this chapter is that each of us is profoundly different and this is why human relationships and society are so rich. It is frequently irrelevant if these differences are due to our genes or our environment or our experiences but this does not imply that it is therefore unnecessary to know what the relative contribution might be. This is perhaps most true in the field of human intelligence because understanding the relative role of education (perhaps the most profound environmental influence on intelligence) and gene differences could lead to a far deeper understanding of how best to design education to benefit all people to the greatest extent. This is surely a great and significant social and scientific challenge and so in the following chapter we will look at the controversial and difficult relationship of genes and intelligence.

Intelligence and the four-way tangle

8

All of us believe we can recognize intelligence but I think it would take each one of us a good deal of time to write down exactly what it is that we feel is represented by intelligence and what it means to possess it; indeed, large numbers of books have been written on the topic and even greater numbers of articles in scientific journals. The difficulty in one way is not profound, for intelligence is not a simple 'property' like weight, length, or height that can be estimated on a scale or with a balance, and so it cannot have a simple and objective measure. Many features of human existence are similarly insubstantial—beauty, friendship, and happiness—whose existence we all recognize, even if we would argue over their definition.

If the measure and properties of intelligence generate much discussion and argument, then the influence of gene differences on intelligence comfortably exceeds even this degree of dissension. The central question we seek to answer is the extent to which differences in intelligence arise from gene differences or from the environment. The answer to this highly controversial question is, however, dependent in part on the solution to the measurement problem and this has been equally controversial. Compounding these two controversies have been further battles over the social implications of measuring intelligence and, most bloody battle of all, of the relationship of intelligence and 'race'.

The battle lines have ultimately been drawn around the interpretation of some experiments using both twin studies and twins adopted at birth (I discussed the rationale of these types of study in Chapter 2). A whole range of studies have shown that gene differences have a moderate effect on our intelligence and that somewhere from 30 to 70 per cent of the variation in levels of intelligence of different human beings derive from the accident of the gene differences they have inherited.

What is so unreasonable about the results of these experiments that have resulted in such controversy? Many different types of studies show that there is a moderate effect of gene differences on many different facets of human behaviour and on the function of the brain (I discuss this topic further in Chapter 6). Why should intelligence not be similarly influenced? At the heart of the arguments that raged from the early 1960s, and still continue to the present day, are several different strands that need to be dissected out so that we can make some sort of assessment of what might be the truth behind the controversies.

The first strand is the measurement of intelligence, and the uses and abuses of the commonest measure, the IQ test; the second strand is the consequence of the gene's influence; the third strand is the potential basis of the gene's influence over intelligence; and the last strand is the possibility that genes not only influence individual differences in intelligence, they may also influence differences in intelligence of different groups of humans. These strands are inextricably tangled together in a four-way tangled knot and resolving one from the other is what I'm going to try and do. I will have to be careful because contained in each strand are many issues that are in themselves important but do not address the primary goal—to see if we can agree or disagree with the relatively cautious statement that I made earlier—that 30–70 per cent of differences in human intelligence are due to gene differences.

■ The first strand—IQ: measure or mismeasure?

Before scientists measure something, they need to define what it is that they wish to measure: this is fundamental to the practice of science. The first problem of measuring intelligence, then, is what is it that we are measuring? Speed of response? Ability to memorize? Ability to calculate? Deduction? Identify pattern? Manipulate language? Recall yesterday's lessons? Even assuming you can answer the question to your own satisfaction, what makes you believe your answer would be acceptable to someone else? For example, I am a scientist and so deductive reasoning—the ability to draw conclusions from complicated masses of observations—is an important part of my work, and high ability in this I would inevitably define as being a central component of high intelligence.

In contrast, a designer may need a developed ability to handle spatial relation-ships—to rotate objects in space or invert, match, dissociate—and so these may well be the more important components of a designer's intelligence.

Could it be that we are thinking about intelligence in the wrong way? Perhaps an intelligent person simply has a brain that processes the electrical impulses, the most basic actions of brain activity, at a higher rate than less-intelligent people. Perhaps intelligence can be related to a general property of the brain rather than some manifestation of this property, such as the particular abilities of the scientist or the designer? At present, there is no evidence in support of this idea: as far as we can tell, an intelligent person's brain seems identical to that of a less-intelligent person in every physical respect.

If you were to think that the difficulty of the definition of intelligence would seem both obvious and essentially an intractable problem, then I have to say that, at this moment, you would be quite correct. So how can we claim to measure an intelligence we cannot define? The approach that has been developed over the past 100 years or so is based on an acceptance that there is no simple, unified definition of intelligence that can be simply measured: instead, tests have been devised that can be used to probe intelligence as a set of abilities appropriate for a person's age. The reasoning behind this is that children, for example, develop intellectual skills as they grow older; if one could develop tests that could be answered by most children at some specified age, but not when younger, then one could judge a child by relative ability—a child would be more intelligent than average if at the age of 6 he or she could solve problems that were generally solved by most children at age 7 but more rarely solved by 6 year olds. Thus, this child would have a test age of 7 but an actual age of 6. A simple measure of intelligence could then be a comparison of mental age with actual age and this is exactly what is at the heart of the intelligence tests known as IQ tests. IQ stands for intelligence quotient and is defined as test age divided by actual age all multiplied by 100; our child would have an IQ of $7 \div 6$, which is 1.17, multiplied by 100 to give 117. If the child was of average intelligence, of course, then he or she would have a test age of 6, an actual age of 6; $6 \div 6 = 1$, times $100 = 100$. 100 is, indeed, by IQ testing, the midpoint of the range of intelligence.

There are several key points embedded in arguments over IQ and intelli-gence. First is the idea that IQ is a description of intelligence but not necessarily an objective measure. In particular, the tests work because the questions are deliberately chosen to give the right midpoint—they are selected by outcome: if the questions do not achieve this, they are replaced with new ones. IQ tests contain a variety of questions that test arithmetical, visual, observational, and inferential skills and the questions have been developed over the years, starting back in the early twentieth century. Individual questions have been retained or rejected based on the desired outcome of the test—a midpoint of 100. Clearly, IQ tests cannot be considered 'objective' in the sense that my weight is objective, and they have been heavily criticized on this basis. I am not convinced that objectivity matters in some respects; it is perhaps sufficient to see IQ as a measure that is related to intelligence but not necessarily identical. In a nutshell, intelligent people may have high or low IQs but, on average, intelligent people have high IQs.

A second point, which is frequently overlooked in the controversies surrounding the result of genetic studies, is that it is IQ and not 'intelligence' that is tested in twin studies. This is perhaps a key point, for it means that even if IQ is an imperfect measure of intelligence we can still discuss the influences gene differences might have on IQ, without addressing the thornier problem of the relationship of IQ to intelligence.

Perhaps the best way of illustrating this is to think of a scientist who is studying weight in human beings. Most scientists would simply weigh their test subjects and obtain an objective measure, but imagine instead that some crazy scientist wanted to determine weight by measuring the amount of food the test subjects consumed. This is not completely ridiculous because there is a relationship between food intake and weight, but it is not necessarily direct; my guess is that provided enough individuals were studied, the results of the two approaches would not be that dissimilar. In exactly the same way, IQ is a reasonable measure of intelligence when applied to enough individuals.

The counter attacks on twin studies of IQ have been based on two approaches: potential flaws in twin studies and the basis of the IQ test. The criticism that there are flaws in twin studies is focused around the fundamental properties of the powerful approach of studying identical twins that have been separated by adoption; if you recall, this approach uses identical twins that have been independently adopted by different parents to test the effect of different environments on the same set of identical gene differences. Objectors have raised the important point (briefly touched on in Chapter 2) that several influences conspire to make the environment of identical twins more similar than that of non-identical twins; for example, there is some evidence that identical twins are more likely to be adopted into similar homes. If this was the case, then the reasons separated twins have similar IQs could be due either to identical gene differences or to similar environmental influences. The response to this objection required much complex re-analysis of experiments and to my mind the balance of the evidence is now that even though adopted identical twins probably do have a greater similarity of environment than might have been expected, the effects of this are quite slight and cannot explain the similarity of IQ—the experiments withstand the counter attack.

The second line of attack was on the IQ test itself and was based on the view that IQ is not a measure of intelligence. I have already discussed this and as a consequence I believe this attack is misguided in as much as we do not have to view IQ as a direct measure of intelligence. There are numerous features of IQ tests that are difficult to reconcile with them being simple and direct measures of intelligence. For example, there seems to be a relationship between increased IQ and social class, and there has been a significant increase in IQ over the past few decades—a very curious development because one would be forced to believe that children are becoming cleverer. But perhaps this is simply an indicator that children can be coached to 'pass' IQ tests! There is every reason to suppose that these facts indicate flaws in IQ testing as an objective measure of intelligence, but this is missing the point: the tests need not be fully objective to show the effects of gene differences.

What I would like to propose now is that we deliberately ignore these issues and make an assumption that the twin studies are correct, because I want to set us free to explore the implications of the relationship of gene differences to intelligence. Perhaps over the next few years or generations, scientists will indeed develop a general measure of intelligence but I suspect that the human brain is so complex that such an idea is unlikely to be realizable; we are attempting to measure the immeasurable. So let us state our starting position: what I am going to suggest is that we simply work on the assumption that IQ is a measure of intelligence and that there is a relationship between gene differences and IQ, such that about 30–70 per cent of the differences in IQ are caused by such changes. By ignoring the controversy, I do not want to pretend it does not exist: but by ignoring it we will be able to explore the consequences of a relationship between gene differences and intelligence and move on to think about the next strand of the tangled knot.

■ The second strand—why waste time and be educated?

A common response to IQ research is the belief that if gene differences define intelligence there is no point trying to educate or to be educated—the genes have done it all and there is nothing further that can be done. This view is totally wrong: if gene differences accounted for, say, 50 per cent of differences in IQ, then the remaining 50 per cent must be accounted for by non-gene effects, by environment. This is not a case of the iron fist of gene determinism; genes are contributing, not defining. Education, reading, practice, parental support—all will make a contribution equal to your genes in defining the intelligence you will have.

The idea that both genes and environment mould intelligence makes a great deal of intuitive sense; how many parents will say of one of their children that they are not academic compared with their brother or sister? Are they saying they have failed to educate one and not the other, or are they accepting that where intelligence is concerned there is a difference of starting point for the two children? Often this sort of comment is accompanied by the observation that the other sib is good in other forms of intelligence—design, sociability, or whatever, and what is being recognized here is that there is a landscape of different intelligences formed by genes that are nurtured by our environment and that the product of both is the garden of intelligence.

The effects of environment on intelligence tend to be forgotten, which also contributes to the fear that knowledge of gene differences will make us individually predictable—predictably intelligent or stupid—and, in the wrong hands, this information can be used to stigmatize individuals or discriminate against the unintelligent. In this case, the central fear is really focused on IQ not genes. IQ is widely used as a component of job selection, for example, but the evidence that having a high IQ will make you successful at a job is quite equivocal; research suggests that IQ is a predictor of job success in about half of cases. Perhaps the easiest way of making this point is to suggest that if you were to take two groups of children, one of IQ 130 and one of IQ 70, and follow them through life, you would find that a greater proportion of people in the higher IQ group would have had financial and job success compared with the lower IQ group. Some high IQ people would have had total lack of success, however, and some low IQ individuals would have had spectacular success. In short, on average, having a higher IQ is a reasonable predictor of ability, but only on average.

The word 'average' is again critical because it means that we cannot say for any selected individual that possession of an IQ of 130 assures success, we can say only that there is an increased chance of success: it is not a prediction of an individual outcome and never will be. Environment, the complexity of the relationship of intelligence to success, and the variable definition of success will all conspire to make us and keep us resolutely immune to individual prediction.

The final point to be made here is the fear of the immutability of the gene's effect—the fear of 'my genes will make their contribution and no matter what I do, they will contribute'. Up to a point, this may be true, but does not always have to be so; some gene differences cause effects that are reversible or modifiable. A clear example of this are individuals with a relatively uncommon single gene difference in a protein called PAH and as a consequence are very sensitive to the amount of a chemical called phenylalanine in food. If untreated, the condition causes mental retardation but it is treatable by a special, low-phenylalanine, diet (I discussed this condition in more detail in Chapter 6). Another way of looking at this is that manipulating their environment—the appropriate diet when they are young—can enhance the potential IQ of these individuals. This is, of course, an unusual case, but perhaps if we understood more of the gene differences contributing to IQ we could identify and eliminate or enhance many different environmental/gene interactions and so influence IQ directly and positively. Precisely by discovering the genetic contribution to IQ we might increase the power of environment to influence ultimate IQ and reduce the gene's contribution.

If we are ever to discover whether we can influence the effect of gene differences, then we must first identify the individual genes that contribute towards our intelligence. What is the chance of this? This is the next strand of the tangled knot.

■ The third strand—many genes make light work?

We know from the twin studies that an unknown number of gene differences contribute towards IQ—it is certainly not one (patterns of inheritance of high or low IQ through families are incompatible with this possibility), it could be tens, hundreds, or thousands. The fact that we don't know how many genes are involved, however, does not stop the search for individual genes using linkage analysis (a method I described in Chapter 2), but up to now, with one or two false alarms, no gene difference has been identified.

This type of research looks deceptively simple and often, because of the great interest in this area of science, results are prominently displayed by the media. The danger in such publicity is that the experiments need to be treated with great caution because there are many pitfalls for the unwary; trying to assess if the science is actually correct or not often involves delving deeply into technical details that are not easily accessible to most non-scientists—indeed, many non-geneticists would have difficulty in dissecting out the weaknesses.

I can illustrate this by considering one piece of research that featured in a UK television documentary and that was subsequently published in a scientific journal: the research seemed to show that differences within a signalling protein called IGF2R contributed to a high IQ. The researchers followed a guess: previous linkage analysis had suggested that a particular region of DNA might contain a gene difference linked to IQ. They guessed it could be the IGF2R gene, which they knew was located in the region, and examined it to see if they could find a change; they found one (by the way, they did not know if the change caused a concomitant change in the IGF2R protein) and they then carried out two case/control studies.

In the first instance, the cases were children with an average IQ of 135 and the 'controls' were children with the average IQ of 103. In the second experiment, they used children of IQ greater than 160 compared with children with average IQ of 101. The researchers compared how often the IGF2R change was found in the cases and the controls, and an imbalance was indeed found. It was not an absolute relationship; the DNA difference was simply more common in, but not exclusive to, the high-IQ group, suggesting that it was one of many changes that could contribute to high IQ. The assumption was that those cases with high IQ but without a difference within the IGF2R gene would have differences within some other gene instead.

The experiment looks convincing but contains fundamental problems that are revealed only by a close scrutiny: there are two main problems in the work. First, the case/control structure is generally held to be unreliable for all the reasons I have discussed earlier. The difference between the high-IQ and lower-IQ individuals might simply reflect the differences in the IGF2R gene present in the population in general and may not be associated with high IQ at all. Case/control experiments are increasingly regarded as flawed and the much more reliable sib-pair analysis is replacing it as the preferred experimental approach—a situation that will alter as our understanding of the relationship of DNA differences within and between groups becomes clearer.

The second objection to this experiment is that an alternative explanation is possible. What if a gene—call it gene X—near the IGF2R gene contains a change and it is this and not differences in IGF2R that contributes to high IQ? How could this be so? Let us think about the origin of the 'intelligence' difference in gene X; it could have occurred by a copying error or DNA damage, but it must have happened in one particular individual. What if, in this one individual, IGF2R was also different—a not unreasonable possibility because every human contains many DNA differences? Now let's follow gene X's change as it goes down the generations and it becomes common, because it makes clever people and clever people survive and have lots of children. Many generations later, gene X is present preferentially in people with high IQ, but what has happened to the original change in the IGF2R gene? It is still there in most cases, because it is so close to gene X that there would be no easy way of dissociating the two. This means that if we look at the IGF2R change in high-IQ people, we are making a profound mistake—right region, wrong gene!

Personally, I am not in the slightest convinced by these experiments: the science is simply not robust enough to support the conclusion that IGF2R contributes to high IQ. What are we to make of this sort of research? Not a great deal, I'm afraid. In fact, I'm fairly pessimistic about IQ research in general because I feel that there are two major difficulties with it.

First, IQ is the product of the massively complex human brain and so could potentially be modulated by literally thousands of genes, making the individual contribution of any one gene too small to be identified. It is possible, of course (and this is the holy grail of IQ researchers), that some gene differences will have a more significant impact—the 'genes of major effect'—and will be detectable in properly constructed studies. But it is an article of faith that this will be so, not of scientific fact.

The second difficulty is that even if a region with a gene of major effect can be located, there remains the task of identifying the specific gene involved. Unlike diabetes, where we know a great deal about the proteins and tissues involved in modulating sugar levels (the 'pathway' of sugar control), we have not the faintest idea of the equivalent pathway of human intelligence. How will we be able to distinguish the causative, high-IQ, difference from all of the other irrelevant alterations that occur within the many other genes in the region?

I am fairly sure that my opinion would change if it became possible to identify some physical differences between high-IQ and low-IQ brains—a different speed of nerve impulse, a different organization—because this would give me the hope that relative few gene differences might be involved in modulating these physical changes. But at present I am unaware of any generally recognized differences and so remain pessimistic. That being said, I would once again be delighted to be proved wrong.

This pessimism does not alter my conviction that gene differences have an influence on IQ and intelligence, even if we will not easily be able to identify the specific genes involved. What I would like to do now is to pass on to the last and most difficult of the strands of the four-fold knot and that is the proposition that because human populations contain disparate gene differences and gene differences contribute to IQ it necessarily follows that different groups of humans may not have the same average IQs: I want to turn to the difficult area of 'race', IQ, and genes.

■ The fourth strand—groups and impossible experiments

I have discussed IQ and also 'race': neither is an easy area, both having a long history of controversy and abuse. What if we put both together? The root of the problem is reasonably simple: IQ tests in the US show that the average IQ of black Americans is 90, 10 points lower than the average IQ of 100 for white Americans, who in turn are 10 points lower than people of Asian origin, who have an average of 110. Huge arguments have raged about the validity of the IQ tests (which were mainly based on the scholastic achievements of white Americans) when applied across cultures, and exactly what was being measured: is it a difference in IQ or could it be a difference in test aptitude, to name but one alternative possibility? These arguments are still unresolved and are, of course, of tremendous importance to education programmes and strategies.

I am not going to try and enter into this part of the argument because I am not expert in IQ testing and an inexpert opinion will add little new insight. But I am an expert in genes and people, and so I do want to enter the part of the argument that suggests that the apparent differences in average IQ are due to gene differences.

I have established that gene differences contribute to changes in IQ; if you are still not happy with this view then, once again, just accept it as being true for the sake of the argument. So, the argument goes on, if individual variation in IQ is caused by gene differences, then differences in average IQ between groups must similarly be caused by gene differences. It is a seductively simple view and the conclusion is wrong, but, like much of science, this clear statement actually hides some rather difficult problems.

Discussing a perceived difference between human groups is rarely easy; history alone is sufficient to ensure this. So I want to do the same as many other geneticists have done (the idea for this came from a textbook written by Dan Hartel and Elizabeth Jones that I use for my student teaching) and instead of talking about different groups of humans, I want to talk about some-thing less inflammatory and less dangerous. The question we are seeking to answer is simple: how do we know that some difference between two groups of the same species of animals is caused by differences in genes? Let's talk about chickens!

Eggs, and farmers Steve and Ann

The topic of chickens, I hope, is free of history and emotion (I am sure I will be proved wrong on this!). Let's say that there are two nearby poultry farms—farmer Steve's and farmer Ann's. Farmer Steve's chickens produce, on average, 150 eggs each year, whereas farmer Ann's produce, on average, 200, and this situation has been going on for many years. How could we show that the difference between the chickens was due to gene differences? Before we go further we need to think what the alternative possibilities are. There are only two. Either Ann's chickens have gene differences that make them better egg layers, or their environment—the feed, chicken houses, warmth, light, or whatever—is superior: gene differences or environment, just as I have discussed many times before. The actual difference we are looking at is also quite straightforward; bear in mind that for both sets of chickens this is the average figure, so Steve's might lay, in the worst case, just 10 eggs and in the best 220—a range of 210—and perhaps Ann's at worst lay 130 and at best 230—a range of just 100. This clearly means that the best of Steve's chickens may actually lay more than the worst of Ann's.

Is there any way that we can manipulate and analyse these numbers to find out if gene differences are underlying the difference? No, no amount of mathematical analysis can do this. Even if you could show that gene differences accounted for the narrowness or broadness of the ranges, you still would not know whether the difference between Steve's and Ann's chickens were gene differences or environment.

So, the first clear point is that arguing over the numbers is a waste of time because it tells you nothing of relevance. What we need to do is an experiment. In all science you do experiments to make observations that can be used to test a hypothesis: if the observations support the hypothesis, then you have a reasonable chance your hypothesis is correct; if they do not, you know your hypothesis is wrong. So the starting hypothesis we are going to test is that the difference between the chickens is due to the different environment of the two farms. If this hypothesis is correct, then the experiment I have to carry out is to swap some of Ann's chickens for Steve's.

If Steve's chickens now start to produce 200 eggs on average (where they were producing 150 on their own farm), we could say that the environment of Ann's farm is contributing to the success of egg-laying. Conversely we would expect Ann's previously good egg layers to now produce a paltry 150 eggs on average. It would be important to swap a good number of chickens; remember the range of eggs per chicken overlaps so we need sufficient numbers so we do not have to worry about the possibility that we, by chance, selected the worst of Ann's to go to Steve and the best of Steve's to go to Ann. So, we do the experiment and we wait 2 years and count eggs. At the end of this time we find that Steve's transferred chickens still lay 150 on average and Ann's a good 200 on average. Wonderful, we have tested the hypothesis and proved it wrong—it is not the environment that causes the difference, so it must be gene differences.

We have, of course, to be careful to check that the chickens that were not swapped continue to lay at the appropriate rate. What if the farms changed food suppliers at the start of the experiment and suddenly all of Ann's original chickens started laying 150? Provided we do this important check, we can be reasonably confident in the result: the difference is in genes, not environment.

A wonderfully easy experiment one might imagine—simple, clear, and definitive: but it is not at all, because all sorts of hidden possibilities lurk in the chicken sheds. Let's look more carefully: what if Ann's chickens are white and Steve's are brown. When the chickens are swapped between farms, whites and browns are for the first time mixed together. Perhaps, having lived exclusively with members of their own colour for years, they start fighting furiously and the minority chickens, brown or white, amongst the numerically superior chickens of the other colour are continually picked on, allowed to feed last, and generally have a poor deal. After a year or so, they would be a sorry-looking collection of feathers and bones with few eggs and not much prospect for better conditions. Under these circumstances we clearly could not conclude that it was the environmental effects of housing or food that resulted in inferior egg-laying.

Surely it's impossible for this sort of consideration to result in more eggs being laid? Unfortunately, it could: what if Ann's champion layers are so fit and healthy, and strong (hence their wonderful egg-laying skills) that they are too tough to be picked on by the weakly stock in Steve's chicken coop? What if they actually fight back and get even more food than Steve's own chickens and so stay as good layers even in a poor environment. To the experimenter, it would look like a genetic effect—high egg-laying being independent of coop.

This is getting complicated: what we need to do is repeat the experiment but keep the chickens separate so they cannot fight but do get the same food and conditions. So we could build a copy of each of the coops on each farm and put the transferred chickens into the copy. Naturally it had better be an identical copy or else the results could be confused again, but this should not be impossible to arrange. We can now define the more definitive experiment: we will transfer the chickens into an identical but separate environment. If they lay differently from their home environment, the egg-laying average is determined by environment; if they lay the same average number of eggs, then this must determined by gene differences. A nice experiment, but we have missed one more possibility: what if it is the collective behaviour of the entire flock that matters—the way they eat or roost, or whatever? Perhaps by separating a smaller number in a new shed, they will not have the same collective behaviour and so will not acquire the good or bad eating habits?

This is where I want to stop; obviously this is getting more and more complex and the experiments more and more difficult to construct, but we are gradually getting towards a definitive set of possible experiments.

As for chickens, so for IQ

It is time to take a break from chickens and to think how we could redesign our experiments to carry out a similar investigation to test the basis of the difference in average IQ between White, Black, and Asiatic; the competing hypotheses we need to test is that either the difference is due to genes or to environment.

What is environment in this case? Almost everything related to growing up, which is almost everything—the effects of upbringing, nutrition, housing, schooling, parental occupation, TV, friends, colleagues, culture, peer group and many, many other factors. Testing each of these in turn is surely going to be incredibly difficult but it's possible to design quite a simple experiment: simply swap a large number of White, Black, and Asiatic children at a very early age and see if the average IQ changes when they are brought up in a household of a different group. Hold on: what about the white chicken/brown chicken problem? How do we allow for the fact that a White child in a Black household might be brought up differently from their adopted Black sibs? Oh this is easy, just change the children by surgery so they have exactly the same colour and other characteristic features of their adopted parents. By now you might be beginning to realize I am not being very serious about this. No one would ever contemplate such an experiment and no one would ever be allowed to carry out; it is an absurdity to even consider such a gross violation of humanity.

We have reached a simple conclusion: not only do we have no way of analysing the IQ of groups to tell us about gene or environmental differences, we also have no obvious direct experimental way of determining the effects of either. Does this mean we will never know? Perhaps not, for two different reasons. Our society is changing and social mobility is far more pronounced than it was even 30 years ago. If we take the US, for example, there are many Afro-American parents who are now solidly in the upper reaches of social and economic success. Their children should face fewer of the economic difficulties facing many disadvantaged black communities and it could be that studying the intelligence of these people is a way of finding out if the obvious environmental factor of economic wealth, with its potential advantages in nutrition, education, and housing, is important in contributing to intelligence. Analyses of this kind have indeed been carried out and show some signs of detecting an influence of the more supportive environment. This approach, however, addresses only part of the problem and I suspect tells us what we already know—that IQ is affected by environment.

I want to return to my chickens again because there is a completely different way of studying the difference between farmer Ann's and farmer Steve's chickens. Let's once again hypothesize that gene differences cause the difference between the flocks. Now let's take a group from each farm and breed them together and look at the resulting 'hybrid' chickens. To avoid obvious environmental effects, let's take all the eggs and make sure half are brought up in Steve's coops and half by Ann's. What we should see in this experiment is quite difficult to predict. If the number of eggs is determined by many gene differences (we suspect, by the way, that this is indeed so), then the chickens that result from the cross will have a mixture of gene differences—some high-laying, some low-laying—and they may consequently lay an intermediate number, let's say 175 on average. We hope both Steve's and Ann's coop groups lay the same average—if not, we have a really hard experiment on our hands; but let's assume that they do.

Now let's take the 'hybrids' and breed half back to Steve's original chickens and half back to Ann's. In each case the original chicken introduces a set of gene differences for either high or low egg numbers. The chicks resulting from the breeding with Ann's high-laying chickens should, therefore, be given additional 'high' gene differences compared with those of their 'hybrid' parent and so should lay a higher number of eggs—say an average of 185—whereas the 'hybrids' bred with Steve's poor layers should be given additional 'low' gene differences and lay less eggs—say 160. An experiment like this is good evidence that gene differences underpin a complicated character; it was in fact used many years ago to show that the behaviour of different mouse strains is conditioned by differences within many genes (as I discussed in Chapter 7).

Can this experiment be done in humans? No, of course it cannot be done in a laboratory, but it is being done naturally and without any intervention on our part. It is delightful feature of our society that we are becoming more mixed—the word here is, of course, multiracial, but I can hardly use this having spent so much time saying that human 'races' don't exist. Mixed marriages are occurring more and more frequently and it is possible that over the next generations the children of these marriages can be tested for IQ and we can see, as they in turn marry and have children, what this does to the average IQ. I am not at all clear that this 'experiment' will be analysable but it is certainly another way to study the problem of group differences.

Personally I have an altogether different hope: it is that the example of mixed marriages will lead us to accept that differences do not matter and that groups are no more than a superficial and artificial way of dividing human beings—that the differences are, quite literally, just skin deep. Perhaps we simply will never bother to analyse these natural experiments because the question we are seeking to answer will have ceased to be interesting.

Do we want to know the answer?

If you have been following this carefully you will perhaps by now have spotted why this particular discussion is fraught with immediate danger. I have discussed why the simplistic view of IQ and gene differences is not supported by experiment and probably never can be. This is not the same as saying that no group differences are caused by genes; we simply cannot tell. It is perfectly possible that group differences are partially determined by gene differences, and as a society we have to be prepared for this difficult possibility because it will provide an opening to racists the world over to continue to argue that one race is inherently superior to another.

One response to this difficult possibility is to argue, as many have, that society should stop all research into the problem; this would be tempting but is probably unrealistic. Research into the genetics of IQ within all humans is important because it could lead to a better understanding of how we should educate people. Just as many have argued that genetic analysis might allow us to choose 'the right drug for the right patient', so genetic analysis may allow the 'right education for the right student', a point to which I will return shortly.

Once information about a contributing gene difference is available, it would be inconceivable that its potential benefit was not made available to all groups within our societies and any 'racial' difference in frequency of gene changes will become rapidly known. The only way society could prevent such a possibility would be to stop all research into the effects of gene differences on intelligence and even, in all likelihood, on some forms of behaviour. Such a decision would inflict very substantial damage on the efforts to develop treatments for many forms of mental retardation and would be an appallingly high cost to pay.

I firmly hold the view that this type of research should not be stopped. Certainly it is imperative that it be of unimpeachable quality, which would immediately eliminate some of it, but the real defence we should mount is based on a purely social consideration—that to discriminate against an individual based on group averages is an absurdity and a nonsense. In every facet of our lives we should simply judge a person's abilities and attributes directly rather than by the colour of their skin or some other indirect characteristic: the approach is surely 'the right person for the right job' rather than 'the right colour for the right job'. Depressingly I am sure that this argument will hold no weight with some people who wish to believe a racist's agenda.

I do not need to know if gene differences cause group differences in IQ because I know I will continue to choose my friends, colleagues, and employees based on their individual behaviour and character, and will care not one jot if they are a card-carrying member of any particular group of human beings; all I wish is for them to be what they are—this is why we are friends or colleagues.

Where on earth are we?

I have discussed the effects of gene differences on IQ on the assumption that IQ is a good measure of intelligence. If you recall, I did this deliberately so that we could pursue some of the implications and misunderstandings that this sort of research produces without having to become bogged down in technical arguments. Naturally the arguments over the relationship of IQ to intelligence have not disappeared and will continue, but will they be resolved? I suspect that they will not in any dramatic way. But I would predict two things.

First, IQ testing will continue to expand to include regions of the human mind that are perhaps considered more the domain of 'personality'—removing the slightly narrow focus on reasoning and deductive ability that present tests have. IQ tests are, of course, only one of a battery of tests that are used to assess intellectual ability and so this prediction is hardly impressive! For my own part, I have always had reservations about the narrowness of the intelligence that an IQ test measures, and we are all aware that very high IQ does not have to be associated with great success in solving the actual problems that confront is in our daily lives. Perhaps a broader set of descriptive measures will ultimately prove more satisfactory.

Second, I would guess that physical and biological measures associated with intelligence may well be discovered. These could be in the physics of how electrical signals pass along nerves or it could be in the way the areas of the brain are organized or connected: I am certain it is unlikely to be a simple and easily observed feature. If we were to make such a discovery, then we could escape from the IQ/intelligence argument altogether, but I really do not think we can be in the slightest degree confident that such physical differences exist. One of the wonderful features of science, of course, is that we can state with certainty what we do not know, but we can never predict what we will know.

My personal position on all of this should be clear to you: I am on the side that believes IQ is indeed a measure of intelligence but I am not at all convinced it is a particularly good measure. I believe that gene differences influence intelligence as measured by IQ and I suspect that there are probably hundreds rather than tens of changes that are relevant. I suspect that, at best, we may identify a few genes of relatively major effect but my guess is that these will be ones that influence very particular aspects of intelligence that are somewhat more easy to describe than general intellectual ability.

Where does this leave us? Not in particularly good shape, I fear. I do not think we are yet close to knowing the true extent to which gene differences contribute to variation in intelligence and we are even further from having the slightest idea of the mechanisms by which they could do this. I would predict that this will remain true for many years to come and it is even possible that we will ultimately conclude that most genes in our bodies can make a contribution and that we will never find genes of major effect.

My discussions so far lead to the conclusion that the headlines and stories of how we will be able to predict and even influence the intelligence of our children by analysing their gene differences are very wide of reality indeed. We have no evidence that such gene differences can or cannot be identified; if we identified such a gene difference it would make a very minor contribution to intelligence. If we knew our child possessed such a change we could not make any statement about his or her IQ but we could make statements about the likely effect on a large group of people who had such a change—they are likely to have a slightly increased IQ. Finally, having a slightly increased IQ is not in itself a useful predictor of the individual future and abilities of our children. There's not much for the headline writer here!

The right education for the right person?

Perhaps the easiest summary of all of the points I have made so far is that it is very unlikely that knowledge of a gene difference will have any dramatic impact on the likely outcome of an individual's life. Does this then mean that this research is pointless? Perhaps not, for two reasons at least. Knowing the identity of a gene difference means we know the protein it produces, which can then become the target for more research, simply because the protein, different or not, must be in a pathway that can modulate, even in a very small way, the way our brain works. This opens new ways of looking at the higher levels of brain function—new pathways, new regions of activity, new functions—and also opens up the development of drugs that might influence the pathways, the so-called 'smart drugs' or cognitive enhancers as they are called technically. It is a very big leap of faith that such drugs can be developed, simply because at present we have no idea of the pathways involved. The complexity of the function of the brain is enormous and it is entirely possible that enhancing one pathway—let's crudely say enhancing the 'IQ pathway'—could have catastrophic effects on some other complicated pathway, such as sociability or speech. Who knows, we could be made smarter but very, very dull to know.

The second reason this work is valuable is that knowing that a person contains a particular gene difference can be an important tool for dissecting the influence of environment. At present we are forced to assume that the environment affects all of us similarly, although we know that this cannot be true from personal experience. We can measure different environments—poverty, housing, reading, parental influence, peer groups—with varying degrees of accuracy, but we have no way of knowing if particular individuals are particularly sensitive, or insensitive, to a given environment. For example, imagine we discover a gene difference that contributes to variation in the ability to respond to verbal information rather than written; knowing this means we can start not only to classify and study the environment (here the relative amounts and type of information coming through speech and writing) but also to classify the response of the two groups who possess or do not possess the gene difference.

Such a study could allow us to change teaching methods that would result in an increased educational success for the children who until now may have suffered from difficulty in verbal understanding. The example is hypothetical but makes my most important point: that a knowledge of gene differences can allow us to alter environments to optimize outcome, rather than be a simple mechanism for crude and stereotypic division of people into 'haves' and 'have nots', 'superior' and 'inferior'.

The desire to apply genetics to human beings, however, has gone hand in hand with exactly this stereotypic behaviour and so it is to this most damaging area of the subject that I now turn—to the 'science' behind the attempts to use genes to define human superiority and inferiority.

■ Eugenics, and genetic superiority and inferiority

The desire to prevent the breeding of the 'genetically unsuitable' is an old, old thread that runs through gene research like a malign demon in an eastern fairy tale. This demon is known as 'eugenics' and this one word encompasses some of the most shameful and awful episodes to which genetics has contributed—the horrors of the extermination camps, the forcible sterilization of minorities, the killing of humans with gene disorders, and the classification of humans into superior and inferior. At its heart, eugenics is based on the view that it should be possible to improve the human species by encouraging 'superior' examples to breed together and discourage or forbid the breeding of 'inferior' examples—in our genetic nightmare, Jean's misfortune is to be classified as inferior.

Is this such an unreasonable idea? 'Yes' and 'no' is the unfortunately complicated answer. No, it is not unreasonable because similar approaches have clearly worked in breeding chickens that lay 250 eggs a year instead of 5, corn with huge heads, cereals with massive yields of grains, and numerous farm animals with high milk yields or muscle bulk. All of these desirable properties are the result of 'selective breeding' carried out by selecting the highest yielding male and female plant or animal, breeding them together, and repeating this selection process in the next generation, and so on for many generations. The results are evident: we have higher yielding crops and more-productive farm animals than any of our forefathers could ever have imagined in their wildest dreams.

What is happening in this series of selections? Many gene differences contribute to the characteristics of farm animals and plants; in each generation there will range from very good to very poor, with respect to whatever is the feature of interest; the best will have more of the gene differences whereas the worst will have fewest or none of these gene differences. Selective breeding means that in each generation only the 'best' pair is selected for further breeding and so, ultimately, only the gene differences that contribute to the desired characteristic will be present in the animals or plants. From now on, the range of differences between their progeny will be much smaller, because the parents only contain the 'good' changes and so all of their progeny will be 'good'.

If selective breeding works for plants and animals, why not for human beings? If IQ is affected by gene differences, what would be the consequence if high-IQ people were exclusively to have children only with other high-IQ individuals? The answer must be that in time such selected high-IQ individuals would exclusively produce as high IQ as the environment of their children would allow. Much the same potential could be realized for any human condition that was reasonably influenced by gene differences—height, weight, personality, and so on.

So what is so wrong with eugenics? In part, a major misunderstanding of the gene's influence, and in part a deeply unscientific acceptance of both the possibility and desirability of classifying features of an individual human life into the 'inferior' or the 'superior': these two misunderstandings combined to produce tragedies of unprecedented scale.

The first misunderstanding was in thinking that the gene differences within two humans with, let's say high IQ for ease of discussion, would necessarily produce a high-IQ child: this is not true. There are many gene differences that must contribute to high IQ and each child gets a chance selection of gene differences from each of their high-IQ parents. Inevitably, some children will get no high-IQ changes, some will get only high changes, and some will get somewhere in between. This implies that the couple's children will have a range of IQs, high to low.

The next misunderstanding arises because we do not know if there is more than one set of gene differences that contribute to high IQ. Indeed, the most likely explanation is that many different combinations of gene differences can similarly result in high IQ. This means that in the extreme case, two high-IQ parents could have two totally different sets of high-IQ gene differences, which would mean none of their children could ever have as good a set of gene differences as either of their parents. Historically, eugenicists were unaware that IQ was formed by many gene differences and that there was no simple relationship of high IQ + high IQ = higher IQ. No doubt, continuing selection over many generations, as has been done for farm animals and plants, would have an effect but only in the long term—tens of generations rather than one.

The early supporters of eugenics were also unaware of the complexities of the interaction of gene differences with environment, at a time when the social extremes of wealth and poverty were the rule rather than the exception. Now, of course, we are well aware that poor diet, bad housing, diseases, and inadequate education can have a very detrimental effect on the physical and intellectual development of children. The eugenicists were even fooled into believing that poverty was a genetic characteristic because it tends to run in families; poor people often have poor children, just as red-haired and blue-eyed parents often have red-haired and blue-eyed children. It was just a short and disastrous step to conclude that the poor were poor because of an inferiority of mind, body, and spirit, which in turn was the responsibility of their genes.

The idea that there was a scientific underpinning to the rigid social divisions of class in English society must have had a powerful attraction to the Victorian founders of eugenics. Foremost amongst which was the Englishman, Francis Galton, who, in 1869, published a book called *Hereditary Genius*, which set out to show that genius ran in families. Galton was to found one of the first eugenic societies and many more followed, all dedicated to the idea that the reproduction of the undesirable in humans was to be suppressed and that of the desirable to be encouraged.

The consequences of the second misunderstanding encapsulated the most repugnant aspect of eugenics—the desire to define 'desirable' or 'undesirable' characteristics. How do we as individuals, or even as a society, decide what is 'desirable' or 'undesirable', 'superior', or 'inferior' in humans? Eugenics provided a scientific framework to justify prejudice of the most direct kind; to the European mind at the turn of the nineteenth century, 'superior' was European and white and 'inferior' was non-European and non-white: 'desirable' was upper class, 'undesirable' working class.

The most terrible abuse of genetics was without doubt the use that the Nazis made of the crude eugenic view of humanity, because, to them, Aryans ('pure Germanic') were superior and non-Aryan individuals, Jews in particular, were genetically inferior and so should be prevented from breeding. The practical impossibility of forcibly sterilizing so many people meant that extermination was the only horrifying alternative, a policy resulting in the deaths in the extermination camps of Europe of over 6 million Jews, countless Poles, Hungarians, Yugoslavs, Serbo-Croats, Russians, Gypsies, homosexuals, dissenters, and the mentally ill.

The science upon which eugenics is based is best understood as an historical anachronism, founded on a flawed understanding of genetics. The abuse of eugenic ideas and the extent to which apparently decent and caring doctors, scientists, and politicians accepted and acted on them should give all of us concern. The banality of evil is perhaps the most profound lesson to emerge from the Holocaust and its perpetrators—the idea that a great evil is a product of small, individually morally insignificant, influences and steps. In this respect, it is very wise to remember that eugenics was ultimately intended to improve the lives of human beings by eliminating poverty and disease, and this goal is shared by virtually all geneticists today. Somehow, the desirability of the end was able to wrap up the means in a protective cloak that hid dreadful abuse and this has been a feature of many well-meaning but ultimately unjustifiable attempts to experiment on human beings. We all, geneticists and non-geneticists, need to be continuously aware of these dangers.

In conclusion

Differences between humans are not only a natural result of being human, it is a critical requirement for our continuing existence—without differences our species would die. The differences between us are partly due to differences within our genes and partly due to our environment and the relative interplay of both of these makes each of us unique and irreplaceable, and also makes each of us respond in a unique way to the challenges of our world. The reality of this far outweighs any attempts to force people into groups and or to stigmatize them based on group attributes and it outstrips simplistic attempts to define skill based on a narrow understanding of gene differences.

Perhaps this way of seeing our 'biological' selves is not so different from the way we think of ourselves in everyday life—each of us responds to the pressures and challenges of our families, jobs, and where we work, live, and play. Each of us responds to these pressures differently based on our individual skills, fears, experience, and deficiencies. Each of us will have defeats and, perhaps, learn from these, and each of us will have victories. In this we are behaving in an exactly parallel manner to the way the biological properties of our bodies respond to and are influenced by the world around us. There is no more to fear about the biology of our genes that there is to fear about our conscious, self-controlled areas of our existence. I hope this is a comforting message to those who may fear the apparent tyranny of our genes; in most cases this fear is misplaced because the influence of our genes is an indivisible, and so perhaps comforting, consequence of our humanity.

Perhaps the greatest achievement of human beings is to have partially transcended our biological selves: technology, medicine, and the societies we have created all work together to insulate us as individuals against the sometimes threatening effects of nature and it is inevitable that our attention has turned to the apparent indivisibility of human lives and gene differences, seeking to alter even this deepest of relationships. It is the extraordinary possibility that we can alter our genes and so escape whatever influence they might have on our lives that I will discuss next.

Can we alter our inheritance?

The lives of both Jeans are inextricably tangled with gene differences but Jeanne Dream is doubly lucky: not only does the knowledge of the differences contribute to her healthy life, she also can escape from the worst excesses by being given new genes—she is privileged because she can escape her Genetic Destiny. Is this science fiction, fact, or merely wishful thinking?

All animals and plants live and die in the shadow of their genes: this is an inescapable fate for all of Earth's living creatures except, perhaps, for human beings. In the past few years it has become increasingly clear that it may be possible to alter the genes in our bodies and so avoid some of their influence. It would seem just a short step to assume that if we can do this for just one gene—say a gene that is altered to cause a disorder—then it must necessarily be just a matter of time before we can alter any gene at will, perhaps enabling us to have any Genetic Destiny we choose: the reality is more complex and difficult than this.

A major part of the difficulty we face is in trying to see into the genetic future and decide what is possible or impossible. These two words hide a curious but complicated and significant question: what exactly do we mean by 'possible' and 'impossible'? This is an important question because here I want to discuss what is possible or impossible about altering genes. But the words can mean different things to different people; as we will see, these small alterations in meaning can have much more dramatic impact on our conclusions than the words alone would imply. I will attempt to avoid this by making a small but important digression—I will try to illustrate why 'possible' and 'impossible' are difficult concepts in genetics and how we can steer a straight path through these difficulties.

■ Possible, impossible, or just plain stupid?

Let me make a series of statements:

1. It is impossible to alter genes using a magic wand.
2. It is possible to alter genes in an unborn baby.
3. It is possible that we will be able to alter the genes in an unborn baby to make him or her more intelligent.
4. It is possible that we will be able alter the same gene in every cell of an adult body.

Each statement contains the word possible and impossible and I am often surprised at the failure of many geneticists to appreciate the care that needs to be taken over their use; nowhere is this difficulty more apparent than in dealings with the media. Let's listen to an imaginary interview between a geneticist and a journalist: it goes like this.

Journalist: Dr Geneticist, is it possible that we will be able to deliberately alter the genes of our unborn children to make them more intelligent than they otherwise would have been?
Geneticist: It is possible, but I think it would be so difficult to make enough alterations to enough genes that it is unlikely to happen.

Next, there's a 25-point headline in the next copy of the *Daily Exciting World*: 'Genius kids on way'. Leading geneticist says 'It is possible' to have super intelligent babies upon demand ...'. Several things have gone wrong in this interview but the key problem is the word 'possible': to a scientist, possible is anything that is not known to be impossible; to a journalist, possible means just what it says—if it is, it can happen; if it can happen, it will happen. The word 'possible' in science is subtle. Let me illustrate this with an absurd question:

Provided there was a dam across the Pacific to stop it falling into the Atlantic and you had a tube to pour the water into space, is it possible or impossible to use a strawberry or a coffee cup to empty the Atlantic Ocean?

I think most of us would reply, bemused by the idiocy of the question, that both are impossible. The question may be idiotic but it is not without value because we are actually giving the same answer for two different reasons. The simple answer concerns using strawberries. These do not hold water and so cannot be used to empty anything, let alone an ocean, and saying this task is impossible is clearly an accurate statement. In contrast, the coffee cup can hold water—albeit a rather small volume—so, in principle, it could be used to transfer water from one place to another, but the task is impossible because we would need to fill the cup too many times. Most people would agree with this answer because it is an impossibly large task, but what we have done here is actually make a judgement without, perhaps, realizing it.

Let's try and explore this by changing the question slightly:

Is it possible or impossible to empty the following volumes of water with a coffee cup: a soup bowl, a bath, a swimming pool, or the Atlantic Ocean?

My answer would be, possible, possible, possible (just), and impossible. The particular interest in the answer is in the transition from possible to possible (just) to impossible. What we are having to do here is assign a simple qualitative description, 'possible' or 'impossible', to steps along a continuous range corresponding to the volumes of liquid in a bowl, a bath, a pool, or an ocean. At some particular point, the qualitative description must change from possible to impossible, but at what point? Would we all agree that at some specific volume the task becomes impossible? Probably not, and so the use of possible and impossible becomes very problematic because we have no simple, objective trigger when applied to a continuum of events.

This problem is faced by scientists on an everyday basis. To go back to the first question, a scientist would certainly say that using a strawberry to empty the Atlantic was impossible but would be forced to say that using the coffee cup to do the same was a possible task. In science, there are unambiguous states of impossibility and possibility, but the continuum in-between is poorly served by simple words—scientists often will use a mathematical estimate based on probability instead.

Even this approach is fraught with difficulty because not all problems are reducible to such analysis and so the way we will overcome these problems is to try and distinguish several words with different meanings within the spectrum. First, an idea can be objectively 'impossible' (the strawberry cannot hold water and so it is impossible it can be used to empty anything). Second, an idea can be 'possible' (a coffee cup can be used to empty a soup bowl), and, finally, it can be 'impossible practically' even though it is theoretically achievable (it is impossible practically to use a coffee cup to empty the Atlantic even though a coffee cup holds water).

Even this last use of 'possible' also needs one last qualification. 'It is possible' does not mean it will happen: so the statement 'it is possible but unlikely that life on Earth will be obliterated by a meteor before anyone reads what I have written' uses the phrase 'possible but unlikely' to describe an accurate statement concerning an event that is vanishingly unlikely to happen, whereas 'it is possible I will mistype a word when writing this book' is a certainty.

So now let's go back to the original four statements and alter them to contain my uses of possible and impossible.

1. It is 'impossible' to alter genes using a magic wand.
2. It is 'possible' to alter genes in an unborn baby.
3. It is 'impossible practically' that we will be able to alter the genes in an unborn baby to make him or her more intelligent.
4. It is 'possible but unlikely' that we will be able to alter the same gene in every cell of an adult body.

I hope that you will see how the meanings have altered and in the chapter that follows, I want to discuss what is possible, possible but unlikely, impossible practically, and impossible in the abilities of scientists to use genes to alter human beings.

■ Gene engineers

To alter human genes requires two very different areas of skill. Obviously, we need to be able to alter DNA on demand, we need to be gene engineers; but because genes work only inside particular cells, we also need to become cell engineers as we have to be able to alter the correct cells on demand. These two skills are, in principle, all we need to possess to alter our inheritance by altering DNA. I will start by discussing the ability to engineer DNA before moving on to engineering cells; finally I will discuss what might be possible by combining both skills.

New genes for old—gene therapy

There are some obvious reasons why it is important to try and put genes back into human beings—it is the ultimate way of 'curing' a disorder caused by a gene difference. For example, cystic fibrosis is caused by a difference to the CFTR gene on chromosome 7, and so if were able to introduce a normal CFTR gene into a patient with cystic fibrosis they should be able to produce normal CFTR protein and so be 'cured' of the disorder. The approach is called gene therapy; it has been tried in humans and unfortunately it doesn't work particularly well—at least not yet.

It would seem rather self-evident that this must be an obvious way of curing gene-based disorders, but I think it is important to understand the limitations and potential of gene therapy before we move on to thinking about how we might carry it out. Cystic fibrosis is at present not a curable disorder but that does not mean it is untreatable; quite the contrary. About half of the children with cystic fibrosis live beyond age 20 and a few live beyond 35. There has been a steady increase in quality of life and this is the background against which the efficacy of gene therapy has to be judged. Quite simply, gene therapy has to result in a better quality of life or a greater life expectancy compared with existing treatments; it is not enough that it works a bit—it must work better or else we are condemning an individual to increased suffering from a less-effective form of treatment.

If we combine this problem with the general prohibition of direct experimentation on human subjects, we are left with a thorny problem: how do we test approaches without testing them on people? The only way of answering this question is to first determine exactly what is it that we need to know before we try therapy in a human being. The nine requirements are quite simply stated but startlingly difficult to study.

- We need to have identified the gene and the protein that are different or missing in the disorder; for cystic fibrosis, this was achieved in 1989.
- Can we identify the switch DNA that controls the genes?
- Can we make a DNA fragment that contains all of the necessary switch DNA and the gene so that it will work correctly when put back into a cell grown in a test tube?
- Will the gene be copied and passed on to daughter cells every time the cells divide?
- How do we put the DNA back into cells of a human being?
- Is the introduced gene copied and does it remain stably in the cells; if not, for how long?
- Does the newly introduced gene make a protein that functions normally?
- Does the newly made protein really 'cure' the disorder and what are the criteria that can be used to judge this objectively?
- Finally, and most critically of all, does the act of introducing new DNA into cells cause any unwanted side-effects?

This is a long list of questions; many have complex answers but fortunately some are already known. We know that just 10 000 rungs of DNA contain the CFTR gene and switch DNA works successfully and produces CFTR protein in cells grown in a test tube. But, even so, we do not know in detail how the CFTR gene is controlled by its gene regulators. Despite this it is reasonable to go to the next question and ask how we might introduce DNA into human cells. This is not easy. In experiments on human cells grown in a test tube, there is a whole variety of methods for doing this; the most efficient and common is a very precise electric shock but, clearly, this is not going to be easy in a living human being! Worse that this, in cystic fibrosis the organ that is invariably affected is the lung, and in some cases the pancreas, and so we have two very different challenges: if we are to alleviate the symptoms of cystic fibrosis we must insert the CFTR gene into cells of all of the organs that are most affected. The lungs are relatively easy to access in living humans but the pancreas is hidden deep inside the body and is far more inaccessible: we face two very different problems.

The experiments to introduce DNA into any cell, accessible or not, in a human body needs to be guided by some general principles and these have already been well established. We know that many viruses can introduce their genes into cells and do so as natural parts of their lives. So one way of getting a new gene into a cell is to add it to a virus—to make a 'hybrid'—and leave it to do the rest. Many viruses cause human diseases, of course, so it is important that harmful genes in a virus are removed from the hybrid: this means that the genes of viruses must be very well understood before we could ever contemplate using one in such experiments. Fortunately a sizeable number of viruses are now well enough understood and can be used for the experiments: each has its own advantages or disadvantages, but four significant problems remain in all cases:

- Not all genes are small enough to fit inside a hybrid.
- Not all hybrids can be guaranteed not to cause disease.
- A hybrid virus may trigger a defence reaction within the body that renders it ineffective.
- Most viruses naturally infect only particular types of cells for reasons that I discussed in relation to the AIDS virus in Chapter 5.

These are central issues for scientists working on gene therapy; they are hard problems to study and hard to alter in the natural lives of viruses. Fortunately there are other ways DNA can be introduced into cells that do not make use of viruses; for example, DNA can be wrapped up in a fatty ball with a rather similar composition to the wall that surrounds all cells. These DNA/fat balls introduce DNA into the cell by spontaneously fusing with the fat layer of the cell when they come into close contact with it. This method has great advantages—no viruses, no four problems, and no unpredictable side-effects—but also one great disadvantage: how do you introduce a sludgy mass of fat balls into a human? In the lung it is possible to make and inhale a fatty ball infusion (a mixture of air and fat) but this is not really an option for introducing DNA into the pancreas.

What happens to the new DNA when it gets into the cell by either route? Once again, we really do not know in detail but there are several possibilities. It depends in part on whether we use fat balls or viruses and, in particular, the type of viruses. Fat balls successfully introduce DNA into the cell. In the best case, this DNA can be incorporated into the cell's own DNA, but in the worst case it is not copied when the cell duplicates its own DNA and so is rapidly lost. Incorporation into the cell's DNA is a somewhat mysterious process that we do not fully understand: we know that somehow the rails of the host cell's DNA ladder are broken and the incoming DNA slipped into the break, which is neatly rejoined. The evidence suggests that this can happen at virtually any location ('randomly' is the correct description) within the 3000 million rungs but, in the best of all cases, the joining can be extremely specific.

Imagine we have put part of the CFTR gene into fat balls—let's say the piece from rung 500 to 700 in the gene. In some cases this CFTR DNA fragment actually finds the CFTR gene in the cell and inserts itself into the exact same place, rungs 500–700, in the DNA—it somehow has 'targeted' itself back to where it belongs. This is extraordinary because in a patient, if we know there is a gene difference in the CFTR DNA somewhere within rungs 500–700, we can introduce an unaltered copy of the DNA and make a perfectly normal gene. We still do not understand how to control the final fate of the DNA introduced into the cell, however, and finding methods to ensure it remains in the cell over many divisions is occupying a great deal of research time at present.

In this respect viruses are very useful because their DNA is, quite naturally, stably maintained inside the cell. Different viruses use different ways to achieve this: the DNA of some viruses remains inside the cell but not inserted into the host's own DNA, and these viruses control the copying of their own DNA independently of that of the cell. Other viruses insert into the DNA of the cell and so replicate as the cell's DNA does—a very useful approach because the virus DNA tends to be more stable in these circumstances.

But there is a sting in the tail of even these helpful viruses; they seem quite as happy to jump into the middle of a gene as they are into a piece of DNA between genes, and this is the most worrying aspect of gene therapy. If the therapy virus by chance inserts itself into a gene within the cell's DNA, then this will almost inevitably result in the destruction of the ability of the gene to make its protein. The effect on the cell may or may not be significant; it depends on exactly which gene has been inactivated in this way.

A much more alarming possibility has already been identified: the therapy virus might interfere with the control of a gene, which can happen when it inserts into the DNA near or within a switch region DNA. In the worst case, the control of the gene can be altered so that encoded protein is produced uncontrollably and this is particularly catastrophic if the gene is needed to control cell division. Geneticists already know from work on viruses in chickens and mice that this event is possible and highly dangerous; the consequence is that cells start to divide continuously and ultimately this causes a cancer. Obviously the prospect of causing cancer whilst attempting to cure a genetic disorder is not anyone's idea of a successful treatment.

Curing cystic fibrosis and clever ways of catching it

These are all serious problems. Nevertheless, despite all of this uncertainty several attempts at gene therapy in humans have been tried. The results are clear: they have not cured the patients, but the attempts have at least shown that DNA can get back into a cell in a human being and that the new gene survives for a time in the cell before it is ultimately lost. Once again, cystic fibrosis and the CFTR gene provide one of the best examples of the current state of the art.

A key feature of cystic fibrosis gene therapy is that it was first carried out in mice in which the CFTR genes had been destroyed by using the gene-engineering technique of transgenesis. This is slightly ironic: gene engineering was used not only to try and cure but also to give the mouse the disease in the first place. The technique of transgenesis is widely used in genetics and so I'll describe it in a little detail before we move on.

In its simplest form, transgenesis is carried out by injecting DNA into a very young embryo when it is no more than a simple ball of cells—in the mouse this is about 3 days after fertilization. The fate of the injected DNA is much as I described for gene-therapy DNA and it can remain separate or insert into the host cell's DNA; the host cell will divide as the embryo grows, carrying its passenger DNA (the 'transgene') with it. The embryo cannot grow successfully outside of a mouse womb, of course, so as soon as it has been injected, it is put back into a female mouse. This process seems a little unlikely, but it is remarkably efficient and can be made even more efficient by using embryo stem cells, which I will discuss towards the end of this chapter (see the section 'Embryo stem cells—endless potential') rather than whole mouse embryos.

Making a mouse with the same gene difference that commonly causes cystic fibrosis in human beings is a little more difficult, because just the one gene in the mouse's DNA needs altering. One way of doing this is to inject into an embryo just a part of the CFTR gene DNA, which is too small to encode the whole CFTR protein. The embryo's cells can be treated in such a way as to force the fragment to insert itself precisely into the natural CFTR gene in the cells; if this is done correctly, it results in the natural gene being replaced with the altered CFTR gene fragment. Naturally this alteration to the mouse's CFTR gene means that it cannot produce the normal protein but, because all animals contain two of every gene, the unaltered gene can still function normally (this is not a good 'model' for the human condition, where the disease is invariably caused by differences within both gene copies).

This failing is easy to remedy in mice because they are quick to breed. The idea is to mate together two mice that have one normal and one engineered CFTR gene, and some of their offspring will have two non-functional CFTR genes. The problem is making only one transgenic mouse in the first place, but this is easily overcome by first mating the mouse with a normal mouse; some of the offspring will contain the engineered CFTR gene and these can be detected by analysing a tiny amount of their DNA, made from just the tip of the tail, so the mouse is not at all affected by the analysis. The mice lacking a functional CFTR gene have many of the symptoms of human cystic fibrosis, so transgenesis made these mice catch the disease.

The researchers then attempted to treat these diseased mice with gene therapy. The first such experiments were carried out using the fatty ball method of introducing the DNA; some DNA was taken up and detected over a period of weeks and there was even an improvement of CFTR function in the lung cells— all very promising indeed. Mice, though, have tiny lungs and are, again, a rather poor 'model' for humans. This immediately raises the key question of whether the same method can work in a human's much larger lung. The only obvious way of answering this was to try actual gene therapy on humans. The results were again promising but a long way away from a 'cure': DNA did seem to get into the cells but stayed for too short a time and there was too little effect on CFTR function to benefit the patient. The result was very disappointing; any effect was much too slight to justify more immediate studies on humans. Consequently the focus of the researchers has rightly returned to examining and improving the many basic processes that therapy must involve and that I mentioned earlier.

An unhappy history

Perhaps surprisingly, CFTR protein was not the earliest target of gene therapy. Attempts to cure the condition that inflicts 'bubble babies' (see Chapter 5) started in 1990, using a virus to carry the ADA gene into cells in the blood of two children. The first indications were that the attempt was a reasonable success in one of the two children: DNA was introduced into the cells and seemed to remain for a period of several years and more. This early success was followed in rapid succession by more than 200 different proposals to treat various gene disorders and over 2000 people were treated. About 70 per cent of the attempts were focused on cancer treatment and the rest on cystic fibrosis (8 per cent) or AIDS treatment (18 per cent); 25 per cent used the fatty ball approach and the rest different viruses.

There have been many fairly wild claims for the potential benefit and success of gene therapy but most of the claims have not really survived the practical reality, which is that DNA is getting into cells for a short period but generally into too few cells for too short a period to alleviate the symptoms or the disorder. This has been recognized as a problem by the bodies that regulate such medical research and over the past few years a much more measured approach has been prevalent: experimentation in animal models of the disease—as in the case for cystic fibrosis—is generally considered to be the essential first step, in preference to premature attempts to move to human trials. Notwithstanding, 1999 witnessed what may have been the first death caused by the side-effects of gene therapy: the details are still contentious and unclear. These sad events have added urgency to reviewing the conduct of these types of experiment. On a happier note just one year later, in 2000, a clear success was achieved in the case of a child with another form of the bubble-baby disorder that I discussed earlier.

The overselling of the prospects of gene therapy has been both sad and damaging because of the false expectations it raises in patients and in the general public; the patients' hopes of a cure are dashed and the public feels confused that once again great breakthroughs vanish only to be re-kindled by the next group to oversell their hopes. Such cycles of overexpectation followed by dampening and reassessment damage the credibility of all geneticists. Why should one believe in the importance of such science when its practitioners cannot even explain a relatively simple truth behind their research?

Perhaps more damaging is the fear that many of the gene-therapy attempts on human subjects have really been little better than experimental studies, carried out even before animal experiments had been conducted; this is unpardonable because there is a wide consensus that human beings should never be used for such a purpose. Defenders of early gene-therapy experiments have used the common excuse that has been applied to many heroic medical interventions—people with genetic disorders cannot expect a cure from any other route and this justifies using experimental treatments. It is a desperately dangerous line of argument precisely because the patients might well have nothing else to lose, and the potential for using their suffering and fear as a lever to obtain consent should require a very high standard of behaviour.

The balance between the interests of the patient and of the experimenter is complicated in this area. The earliest attempts at gene therapy preceded even those on ADA and involved experiments carried out some 10 years earlier, in the early 1980s, using globin genes to cure the genetic disorder of thalassaemia. This disorder is particularly common in the populations bordering the Mediterranean Sea (*thalassa* is the Greek for sea and to the ancient Greeks there was only one 'sea'—the Mediterranean). Its underlying causes are differences within globin genes that result in a failure to make globin proteins—the constituents of the red oxygen-carrying pigment, haemoglobin, within our blood. The state of knowledge in the late 1970s and early 1980s was quite unequivocal: all attempts to reintroduce globin genes into cells, human or not, had failed to yield significant amounts of globin protein. Despite this, attempts were made to introduce a globin gene into two patients, which, quite predictably, failed to produce any beneficial effect.

Why was this experiment allowed? The scientist involved was not granted permission to carry out the experiment in the USA by the local medical committee that reviews human medical research. Rather than wait for scientific knowledge to progress to a level where permission would have been forthcoming (a wait of some 10 years, as it turned out) the experiment was instead performed outside the USA in two countries where perhaps the researchers were less well equipped to understand the scientific reality underpinning the attempts. This was a shameful episode in medical genetics and it is perhaps more interesting to consider the potential motives of the scientist involved; why did he wish to try such a misguided experiment? Of course, we do not know for certain but a general possibility is that if the experiments had worked, they would have been a genuine breakthrough in medical science, the 'cure' of the first genetic disease, for which a Nobel Prize might well have been appropriate.

The lessons we can draw from this history is that, given the tantalizing possibility of glittering prizes has not completely receded, every gene therapist must make enormous efforts to avoid self-interest obscuring the issues of informed consent from patients: the line between self-interest and patient interest has perhaps become hazy in this field of medical science. The more recent experimental failures of gene therapy, when combined with the potential for abuse of patients' rights, triggered the recent re-appraisal of the whole area of this research. What is emerging is a less dramatic, but more realistic, programme that seeks first to answer some of the fundamental questions that I have raised. Only after these basic problems have been resolved can we seriously return to the application of the knowledge to the patient.

Will it really work?

Do all of the difficulties with this research mean that gene therapy will never work in human beings? We are in the realm of opinion and possibility: we simply know too little at present. But my own feeling is that the approach will almost certainly work in a limited number of cases—in particular, where a disease affects cells that can simply be isolated from the body so DNA can be easily introduced. It is likely to be most effective in disorders where a difference within a single gene has a clear effect, which is uncomplicated by environment or other gene differences: such conditions are relatively uncommon.

Perhaps surprisingly it is for cancers, which clearly do not meet this one gene-difference criterion, that gene therapy may prove most immediately important. Cancers may not have a simple origin in a single gene difference— quite the contrary, many genes are different in almost all cancers—but they are open to attack from many different directions, using different biological routes to mount an attack. In many cases, treatment is being attempted by switching on the genes that control the natural cell-death processes that I discussed earlier (Chapter 2); in others, cancer cells are induced to take up a gene whose protein product can be used to kill the cell by drug treatment—so-called suicide approaches.

It is unlikely that a single treatment applicable to all cancer cells will ever be developed, simply because the different cancers have different origins, but cancer gene therapy has already shown some promise when it is used in conjunction with other types of treatment. What is being achieved here is perhaps not gene therapy in the 'pure' sense of correcting a disorder caused by a gene difference; rather it is using genes as a new way of introducing, sensitizing, or directing therapy.

Last, but not least, medical geneticists must face up to the difficulty that not only are many technological aspects of gene therapy unknown, but so also are many of the details of the disorders they are attempting to cure. Perhaps the example of MAO and violent behaviour that I discussed in Chapter 6 will crystallize these problems. The unanswered question in the case of MAO was whether the lack of the protein had caused an alteration to the normal development of the brain; if this was the case, then the alterations would be irreversible, whereas if the alteration of behaviour was a product of the lack of the protein in the adult, then gene therapy to replace the protein might have benefits.

The idea that gene differences might irreversibly influence development suggests that some genetic conditions could be functionally irreversible. Are such early differences really likely to cause diseases in adults? Very much so; for example, there is the substantial body of evidence that this is exactly what might happen in some types of diabetes (Chapter 2). At present, there is little information about which disorders might similarly be conditioned by early events and so we may be attempting to treat the untreatable.

The idea that events in early life may define a genetic disorder raises an obvious point: why not alter the genes in a human baby even before it's born, and so cure all of the resulting abnormalities of development while at the same time ensuring a cure for any adult defects? This is a reasonable scientific possibility because it has been achieved thousands of times in mice and in many ways it is a far easier approach than any attempts at gene therapy in the adult. So I would like to move on and discuss the potential of using 'embryo' gene therapy.

The sins of our fathers

The idea of embryo gene therapy is attractive because all of the mouse experiments show it works. As in many successful transgenesis experiments, DNA can easily be reintroduced into early mouse embryos and subsequently appears in most cells of the resulting baby mouse. The challenge becomes much simpler for the geneticist because the process of introducing the DNA into an embryo is so well understood; in a nutshell, delivering DNA to a few tens of cells is a far easier goal than to many millions, which has to be the goal of adult gene therapy. If gene therapy could be carried out using human embryos, all that really is needed is to identify the DNA regions that are required to control the time and place a gene is used.

How could such gene therapy on human embryos be attempted? The route is fairly straightforward and, technically, there is no reason whatsoever to suppose it would not work. Initially all that is required is to use the methods that have been successfully developed to achieve 'test-tube' babies, which are, in scientific language, the results of *in vitro* (meaning 'in glass') fertilization, or IVF. This starts with sperm and an egg mixed together in a test tube, a sperm fertilizes the egg, which then starts to divide. Even though the cells are still outside the womb, they divide several times and become a ball of a few tens of cells—

exactly as occurs during the normal development of any embryo whether it is fertilized *in vitro* or more naturally. Provided it is placed in its mother's womb soon after this stage, the embryo will go on to grow into a healthy, normal human baby.

The important point here is that for a few crucial rounds of cell division, the embryo is freely accessible in a test tube and so a new copy of a gene can be introduced into it by injection of DNA or infection with a virus. Once this has been done, there is still time for the embryo to be placed inside the mother, whereupon it will grow and develop normally. The embryonic cells—perhaps all, perhaps a proportion of them—will contain the therapy gene and, with luck, this will function as the normal gene should. As I have said, all of this has been achieved many times in mice and all except the injection of DNA has been carried out on human beings: this last step is illegal in many countries.

Why should this approach be illegal? For one very simple reason: there is no way of stopping the introduced DNA getting into the embryonic cells that are going to become the egg or sperm cells in the adult, and this means that not only do we have to consider the effects of the therapy on the baby that is to be born, we also have to consider the effects on the baby's babies, their babies, and so on down the generations—perhaps for as long as there are humans alive on the planet. This is a daunting prospect: how can one predict the future impact of this experiment? Worst still, there is no simple way of testing an embryo to prove no damage has been done by the incoming DNA and the prospect of inadvertently damaging the embryo is extremely disturbing.

Such adverse effects can happen in any type of gene therapy—adult or embryo—but there is a real fear that the complicated process of development would be particularly sensitive to early loss of critical genes, in the worst case producing babies with serious developmental disorders. We know from thousands of mouse experiments that these fears are not misplaced: in about 10 per cent of such experiments, the inserted DNA has an effect on subsequent mouse development (admittedly only when mice are bred to have two copies of the gene). The important point is that young mice can be examined to see what the problem is and, if necessary, can be destroyed before they have suffered too much. Clearly this is not an option in humans, no matter how rare the adverse outcome may be. There is a quite brutally simple question that demands an answer: what do we do with failed human experiments? The answer is, to me, very clear. Because we cannot contemplate inadvertently creating a damaged human life, gene therapy on human embryos should be banned.

This leaves us with two competing imperatives: on the one hand is the imperative to help alleviate suffering, which implies that gene therapy must be used effectively and suggests embryo therapy as a possibility; on the other hand is the imperative to preserve the sanctity of human life. The tension between these two imperatives is obvious and it is unsurprising that the first experiments in embryo gene therapy are in progress. I do not welcome these because it is my view that we do not know enough about what happens to the DNA in an embryo therapy experiment, nor do we have the techniques to establish the location of the DNA and therefore its safety; the possibility of damage to the embryo is unacceptably high. It is a point that I will return to later; certainly there is much room for discussion, but at present, for me, embryo gene therapy is one step too far.

Reality check

The current scientific reality of gene therapy, therefore, is that it really isn't very good: this is not in the slightest bit surprising because geneticists still have a very incomplete knowledge of how genes are controlled and how DNA is moved within a cell. Until there is this understanding, it is likely that gene therapy will remain difficult and ineffective. This does not put gene therapy into the 'impossible' category—far from it because all the science points to the fact that it will work once the fundamental questions are understood. So gene therapy is in the 'possible' category but I think it will be difficult to make it completely routine because of the need to control the way the therapy DNA inserts into the cell's DNA. In theory, ways can probably be found that will give certainty of outcome, but can we be sure that these are achievable?

What I'm implying is that some therapy attempts will end in disaster—a cancer being the most likely one—and so therapy will be justified only where the suffering of not treating the disorder outweighs the risks of treatment. Are you uncomfortable with this idea? Then does it reassure you to point out that many existing treatments for cancer fall into a rather similar category—they are partially effective, they can have severe side-effects, they may even induce new cancers, and they can fail but, above all, they can sometimes cure? The point is that, in most cases, without treatment a cancer patient's life expectancy and quality of life are dramatically reduced; the ambiguity of gene therapy is not novel.

I predict that when historians of science look back on gene-therapy research and the uses geneticists have made of this knowledge, they will see a very ambivalent history. I suspect the judgement will be that we were arrogant and came extremely close to using humans as experimental animals; we knew too little and expected too much, and the expectation of success was used to roll back objections. For these reasons, it is not an area of research that I admire. But I'm convinced that the more measured approach that is now being implemented, based on a deeper understanding of the problems faced by gene therapy, will in the end show it has an important and valuable place in medicine.

The contribution of gene therapy to medical treatment is most probably still in the future, therefore, but a more immediate prospect is the chance of choosing the gene differences for your children by making the best out of what you already have. It is to this area I now turn; it is one where powerful new technologies are pushing back the frontiers of the possible.

■ The best of a (bad) job?

How could we consider choosing at least some of the gene differences our children should receive from us? This is actually not quite so difficult as you might imagine and would not need any gene therapy or engineering or anything other than some natural process that occurs every time a baby is conceived as well as some very clever technology, which already exists. The egg of the mother and the sperm of the father that will make a future baby together contain a combination of gene differences that represent a random selection amounting to half of each parent's DNA differences. This random-half figure means that every fertilized egg from the same pair of parents will have a different combination of gene differences, a different mixture of the parent's complement of differences.

This observation opens an interesting possibility. If there are so many possible combinations of differences, why not simply work out a clever way of choosing only the baby that has the 'right' combination of gene differences from each parent? For instance, if a father has red hair it could be because he has one blond-hair gene difference and one red-haired (this combination produces strawberry blond hair) and so about half of his sperm would contain the blond gene difference and half the red. So if the mother is a blond who wishes that their baby also will have blond hair, all they have to do is to work out some clever way of ensuring that the sperm that fertilizes the mother's egg is one that contains the blond, not the red, gene difference. This is actually not easy because we have no way of testing one DNA molecule, which is all a sperm contains, without destroying it: besides, the parents are going to want to choose much more than hair colour and so will have to look at many gene differences at once.

In the absence of choosing the right sperm or egg to make a baby, how, then, might the parents proceed? Consider this story: a couple want only their best characteristics passed on to their child—no gene differences that might predispose to cancer, heart attacks, or Alzheimer's disease; perhaps a few dozen gene differences must be excluded from their child. So they go to a test-tube baby centre and donate sperms and eggs that are fertilized in a test tube, *in vitro*, just as many fertilizations are carried out nowadays. Each fertilized egg is carefully monitored and allowed to divide to a couple of dozen cells. Then from each embryo a few cells are removed using micromanipulators (tiny needles and tweezers whose minute movements are controlled by hand through a series of simple gears) and the rest of the embryo is then carefully frozen. The cells that have been removed will grow easily in the right test-tube conditions but this happens in a rather disorganized way, and if they were put into their mother's womb they would never turn into a normal embryo. This fact will not matter because in a short time there are enough cells growing in the test tube that they can be broken open and the DNA extracted. More importantly, because the cells are growing in an artificial environment an unlimited amount of DNA can be made; if more DNA is needed, more cells are grown.

Each sample of DNA comes from a separate embryo and so each will have a variable complement of the parent's gene differences; it is an easy experiment and would take a competent gene engineer no more than 4 hours to test each DNA for any differences that the parents wished to include or exclude from their baby. All the gene engineer has to do is identify the embryo DNA that has none, or fewest, of the 'undesirable' gene differences and all, or most, of the 'desirable' ones. Having done this, the chosen embryo, which has been patiently waiting deep-frozen, can be thawed and put into the woman's womb. It will start to grow once again and develop into a baby that will be less likely to suffer from diseases, because it has fewer of the 'undesirable' gene differences than some of its potential brothers and sisters would have had, and more likely to have the 'desirable' characteristics that the parents have selected. In some respects, then, it will be the best of a bad job.

Is this story really possible? Yes, it is and every step has already been carried out successfully on human embryos. This statement might surprise you but the reasons for it are quite straightforward; most are part of the current practices of IVF that have been carried out for more than 20 years—the first 'test-tube baby' was born in 1978. This means that collecting eggs and sperm, carrying out fertilizations, starting to grow embryos, freezing embryos, and putting them into the mother are all routine and there are well over 150 scientific reports of carrying out DNA analysis on cells taken from a fertilized embryo and subsequent successful growth of a baby. In short, the techniques work.

The problem of accumulating numbers

There is once again a serious practical difficulty, however, and that is in numbers. The biological fact is that a woman has a limited supply of eggs in her ovary and so we run into a problem with numbers, and numbers matter badly. The reason for this is that many embryos will inevitably not have the desired gene differences; this is most graphically shown by eye colour. Let's say that the parents want to have a blue-eyed child but the father has brown eyes. He can have a blue-eyed baby provided his own eye-colour gene changes are favourable: he must have one brown-eye gene difference and one blue-eye (this mix is not that uncommon—the brown-eye difference masks the blue-eye one). Half of the embryos that could be made from his sperm and the mother's eggs would have a brown-eye gene difference within them and so these embryos would be no good. But this is actually the best of the situation; the worst is if both mother and father have a brown-eye gene, then only a quarter of the embryos would be blue-eyed.

We have a logistics problem here: the more genes we are trying to avoid passing on to the child, the more embryos are no good. It would take only 10 genes to require that about 1000 embryos would have to be screened before one was found with all 10 genes in their desirable form. Unfortunately a woman produces only a small number of mature eggs at a time. Each month a few eggs are selected from the million or so undeveloped eggs in the ovaries and these start to mature. We do not understand which immature eggs are initially selected, nor why only one or two of the chosen eggs will go on to develop successfully (the others die). So, for the present, until—and if—we can find ways of helping women to make more eggs, we are stuck with small numbers and so incomplete choice.

DNA on a chip—what is possible

A major problem remains in all this: how do we know what gene differences to check for in our couple's DNA? In part, this is easy: just check the mother's and father's genes, find all the differences, and then test the embryos for these genes—not difficult in theory but enormously expensive at present because of the huge amounts of work. Within a few years, however, even this state of affairs will change because geneticists will have a fairly comprehensive idea of all of the common DNA differences in our genes, as well as many of the not-so-common ones and so will be able to test for just these differences. But will researchers be able to test our embryo DNAs for differences down to perhaps the few thousand that this checking would require? We don't even have to worry about it because such an experiment can already be done by using DNA on 'chips', a new and ingenious technique for analysing gene differences (I go into more detail about this in the next section). The potentially revolutionary nature of this approach means that, unusually, I shall deviate from the simplest path through genetics to discuss a little of the technology behind analysing DNA using these clever devices.

The challenge of detecting many different DNA differences in one analysis is enormous: at the extreme, to detect every DNA difference in an individual, all 3000 million rungs would need re-analysing—a new Human Genome Project for every individual. By any current technology, this would be impossibly slow and expensive. Developing a radically new technology that could instantly analyse every rung in an individual's DNA is without doubt the Holy Grail of DNA technology but, like the Grail, it is an illusive prize. Nevertheless, DNA on chips can provide a huge amount of information about the DNA differences in an individual, but only under one initial condition—you must first know what the differences are before you can analyse them.

This is not an unattainable goal; I have extensively discussed some of the huge efforts going into identifying the common gene differences that contribute to the common and not-so-common disorders and the next few years will see a plethora of DNA targets for chip experiments. At the time of writing (2001), more than 21 000 different gene differences have already been identified in just under 1000 genes. This list of genes does not necessarily include all the genes that might be of interest to parents empowered to choose the features of their child, and the limits of parental choice have not been established—a point to which I will return.

Just one difference in three thousand millions?

The task of detecting just the single altered rung amongst the millions might seem impossible, but it is not quite so difficult partly because of the unique nature of the DNA ladder and partly because we know in advance exactly what the DNA difference is that we wish to detect.

The DNA ladder, if you cast your mind back to Chapter 2, can be split into two just by heating it. The unique feature of DNA is that each half ladder will fit back perfectly together to make a whole ladder, because each rung is made of a pair of chemicals—A and T or G and C—with no other permitted pairs. So let's consider a single gene that has a single T–A rung difference such that most of us have the 'normal' rungs AGCTTGA but some of us have AGCATGA. The challenge is not simply detecting the T to A change, equally it's detecting which of the two seven-rung sequences are in an individual's DNA and realizing this is the key to understanding DNA chips; there is a simple way of doing this.

Making DNA is routine in genetics labs all over the world these days, and so to make a DNA fragment with the rungs AGCTTGA is simple; it is also reasonably simple to attach one end of this DNA to a solid surface, often glass. The DNA can be attached as a small dot or 'spot' on the glass and this attached DNA behaves much like a DNA that is not fixed in this curious way. Now let's do a simple experiment and, to make this easy to understand, I am initially simply going to list what we'll do without explaining how it is done.

- Make some human DNA and dissolve it in water.
- Attach to the DNA a chemical that glows red when it is subjected to light of exactly the right colour (a laser light source is used for this).
- Break the DNA into short pieces.
- Boil the DNA so that its two rails break apart, exposing the half rungs.

Now comes the ingenious part ,which needs more explanation: we carefully dip the glass, with our attached DNA as a spot, into the solution of DNA and leave it to soak for several hours at a carefully controlled temperature, after which we remove the glass and wash it in large amounts of water to wash away the DNA that was in the solution. (Remember, the DNA in solution was the one that glowed red when it was illuminated, but the AGCTTGA immobilized on the glass was not modified and hence cannot glow.) Now we shine a laser at the spot on the glass where it was attached and we find that it is glowing bright red! How could this happen? The DNA attached to the glass was never treated to make it glow but it is indeed glowing and the reason for this is quite simple: the DNA on the glass has found its partner DNA rail in the sample of human DNA floating in solution.

This needs a little explanation: the rule that A pairs with T and G with C means that AGCTTGA naturally pairs with TCGAACT, but we made only the first DNA and attached it to glass. When we soaked the glass in the DNA solution, the temperature was designed to allow the DNA to find its partner in the human DNA in solution and make a perfectly normal DNA ladder, with one of the rails capable of glowing red; this is why the 'spot' of DNA now glows red.

The specificity of rung formation is very high, as I've mentioned, with A able to pair only with T and G with C, so, naturally, AGCTTGA can make rungs only with TCGAACT and if there is just one rung different between the two DNAs, they will not be able to make a DNA ladder. This simple fact is how we can make a detector of single-rung differences. To do this, we have to make both AGCTTGA and AGCATGA, and 'spot' them onto glass as two separate spots, just as I have described. If we now take the red-glowing DNA of a person and repeat the experiment, we can see three possible results:

- Only the AGCTTGA spot glows
- Only the AGCATGA spot glows
- Both spots glow

Why? Because if the person contains two copies of the T gene difference only the AGCTTGA spot will find its partner; if the person contains only the A gene difference, then only the AGCATGA can find its partner; and, finally, if the person contains one A difference and one T then both spots will glow.

We have achieved our goal: we have built a detector that can identify a DNA difference in just one DNA region. But the wonderful thing is that we can easily put more and more 'spots' of DNA onto the glass—in fact, we can put many thousands in carefully organized 'arrays' of spots on a surface no more than a few centimetres square. Provided we have computer software that can keep track of what DNA has been spotted where, and which are glowing red and which are not, we can, in one experiment, detect many thousands of DNA differences in a single analysis. The technology behind these 'arrays' is extremely complicated; in one case, it uses similar methods to those required to make computer 'chips' and hence the phrase 'DNA on a chip'. It is by no means a simple task and geneticists still face many practical difficulties before we could make a 'chip' with all of the gene differences that we think are important. But the potential is enormous and, to some large measure, already realized.

To return to the use of this technology for embryo choice, we could, without doubt, use the existing 'chips' to identify the embryo, even if choice might be relatively limited: this much is science fact and not science fiction and we are entering a new world of consumer choice. It is only a short step to realizing that if it is possible to identify gene differences, it must be possible to contemplate identifying gene difference that influence 'normal' features of our existence—intelligence, size, looks, or beauty—because all of these are influenced by gene differences. So let's turn to the idea that we may be able to have the gene differences we desire, rather than the gene differences we have been given, because in doing this we move the potential of gene therapy from the arena of medical treatment to the arena of individual choice and preference—to a new moral and philosophical dimension.

Designer babies—brains, beauty, or both?

I think we would all agree that most people want the best for their children—the best education, the best presents, the best housing, and the best environment in which the child can explore, learn, and develop. If our children become sick, their suffering is ours until we can help them. Given the opportunity, who would not want their child to be born beautiful, clever, strong, healthy, socially popular, and admired? Possession of these attributes would be a great start to a happy and fulfilling life, even if we know that the adult reality may be very different. Well, the evidence suggests that gene differences play a part in determining these features. So why cannot parents choose the gene differences in their child so that they can be born with these advantages? If gene therapy may work for curing a disorder, why not use it for making a baby with more desirable features—a 'designer baby' in other words?

There are really two quite separate issues about designer babies. Could we design them and should we design them? I suspect that far more attention has been paid to the 'should' than the 'could'. So let's try and start by trying to see what is possible, what is impossible practically, and what is impossible. Perhaps the easiest way to start is to list a few of the features that people might seek to build into a designer baby and then to consider what is known about the gene differences that underpin these: sex, hair, skin and eye colour, musculature and strength, good looks or beauty, social ability, intelligence, and, finally, the absence of detrimental conditions. Some of these features I have mentioned before and so I will try and skim rapidly over the contribution of gene differences.

- *Sex*: this is determined by just one gene and if an embryo has a working copy of this gene at an early stage of development, it will become male; if not, the embryo will remain female. Identification of the sex of the baby is already possible.
- *Hair, skin and eye colour*: these features are determined by relatively few gene differences; hair colour and, to an extent, skin colour by differences in the MSHR protein (see Chapter 2) and eye colour by differences in at least two genes. Other gene differences are certainly involved as well, but these are not yet understood.
- *Physique*: the gene differences that contribute to differences in musculature and strength are poorly understood. Twin studies suggest a very substantial influence of gene differences and studies in mice have certainly identified at least one signalling protein, STAT5, which is important in making male mice larger than females. But otherwise there is really little more to go on. One would imagine that differences in physique would be determined by differences in several or many genes as well as by strong environmental influences.

- *Beauty*: thinking about the possible effects of gene differences on good looks or beauty identifies two problems. One of my students recently looked at all of the available information and concluded that even with our relatively limited knowledge of human development, over 80 genes had already been shown to play a part in the development of the brain and head. Identical twins look very similar—hence the name—so, of course, gene differences must be important in defining looks, but geneticists have absolutely no idea of how they contribute to making the unique features of our face. More importantly we have even less idea of how these would make someone handsome or beautiful— twins can be 'identically' ugly or 'identically' beautiful. The second problem is an age-old one, quite literally; there is ample evidence that the ideals of beauty are, in part, socially defined and the ideal of beauty represented in Rubens's voluptuous women is clearly different from the slim, perhaps boyish, figures of women painted by Botticelli. Even if we could agree on a definition of beauty, the complexity of the genetic differences and environmental influences defining the necessary shapes of the body and face would be huge.
- *Personality and behaviour*: the twin studies I discussed in Chapter 7 show the influence of gene differences on our most complex of behaviours, but we don't know how many differences are involved nor the identity of the genes.

Don't invest in designer-baby companies

This leaves the erstwhile designer of babies in very bad shape and facing problems of Himalayan stature. In all but the first two examples—sex and eye, hair, and skin colour—the identity of the genes needed to make designer babies is simply unknown. Similarly, in all but these first two cases, we can be sure that if one gene could be altered the resulting effect in the baby would be extremely small, because one gene difference contributes only a small part of the whole. Furthermore, even if the appropriate gene change was possible, the environment might well negate any effect this difference might have, because in almost all cases the relationship of the effect of a gene difference and the baby's environment—where one can reinforce, negate, or reverse the effect of the other—is simply not known. We could hope to identify more than one gene difference but this would simply increase the complexity of the potential interacting effects.

The outlook for nascent baby-designing companies, therefore, appears bleak; don't invest just yet! Most designer-baby schemes are really in the realm of the 'impossible practically' category simply because there are, in most cases, no genes to design. But let's be wildly optimistic and suppose that in time this will change; we will assume we have many genes to work with. Even so, we would face huge practical difficulties. To alter the way we look would require gene therapy of embryos rather than adults because it is difficult to see how we could use a gene to remould an ugly face into a beautiful one; the shape of the face must be defined, in part, as an embryo develops into a baby and a baby into a child and child into adult. This means that designer babies will definitely require embryo gene therapy.

As I have explained, all types of gene therapy carry a risk of side-effects; if we had the systems to carry out gene therapy (we don't) and if we had the many genes need to make a beautiful face (we don't), then we could introduce these genes into an embryo and make an embryo with a beautiful face—unless a side-effect killed, maimed, or injured it. The simple rules of accumulating risk tell us that the more attempts at gene therapy we make, the greater the likelihood of injury to the embryo. To alter a face might require a dozen therapy genes and increase the risk of an adverse effect dramatically: would this risk be worth the gain? I doubt it—what parent would be impressed by the chance of a physically beautiful child born with cancer or a lovely face but no legs?

In short, designer babies are impossible practically and I suspect will remain so unless we can learn to introduce DNA into cells with complete specificity and safety. We really do not know if this can be done; the research into gene therapy will provide some of the answers but, based on what we know now, it would be very surprising to discover that gene therapy can be carried out with no possibility of adverse reaction. Risk will always remain, therefore, and be greatest for the treatment that requires most steps.

Passing the limits of choice?

The practical impossibility of designer babies has not stopped endless discussion of the morality of their creation—genetics has often been understood with all of the technical rigour that we apply to our appreciation of science fiction. But the discussion of speculative abilities of geneticists should always be welcome because it provides a way of thinking about problems in advance of their existence. The main difficulty with designer babies seems to be with the idea that we are moving from the area of medicine into the area of personal preference. For example, many people, but by no means all, are unhappy with the idea that one could choose a baby's hair colour, which is, after all, just a matter of fashion or personal preference.

Central to much of this debate is the view that treatment of a disorder is morally different from intervention to define a feature that cannot be described as a 'disorder', but 'disorder" is an extraordinarily difficult concept to define. The choice of red hair would appear to be firmly in the category of personal preference but this is not quite so clear cut. Red hair is often a characteristic of people with pale, freckled skin that rarely tans in sunlight. The lack of tanning makes their skin much more sensitive to the damage sunlight inflicts on skin cells and consequently they have a much higher risk of developing the dangerous type of skin cancer called melanoma. In this sense, red hair can actually be considered a 'disorder'.

There is no easy resolution to these arguments, which are not focused solely on designer babies but also apply to more-conventional medical practice; the debate continues and the practical difficulties that I listed earlier would suggest that there is no particular hurry—designer babies are not going to be in the shops this Christmas.

Environment and the end of medicine?

There is enormous public interest in gene therapy and genetic selection and this is often focused on the wonderfully powerful and ingenious technology that geneticists now use. This seductive mixture makes it doubly important to understand that genetic conditions don't have to be treated with any of these high-tech methods. We must not forget the equation, genes + environment = you, which means that changing the environment is an equally logical approach to altering ourselves. Given the relatively modest contribution of gene differences to most variable features of our lives, it is likely that environment has the more profound part to play in moulding our individual features and abilities. Therefore why not modify outcome by manipulating our environment and leaving alone the complexities and difficulties of gene analysis and therapy? Perhaps the best example of such a possibility is that of phenylketonuria, or PKU, which I discussed in Chapter 6. This condition is reasonably 'curable' by reducing the amino acid phenylalanine in the diet—a classic case of altering the environment to counteract the effects of genetics.

Any research into human differences that does not recognize the role of environment is clearly missing a significant part of the equation. The environment, however, is an even more gigantic subject than a paltry few hundreds of thousands of human genes. The key difficulty to date is that attempts to study environmental influences on human lives have been forced to contend with almost complete ignorance of the gene's influence. Researchers have had to study both at the same time, which has clouded understanding. For example, the case of PKU looked at from a slightly different perspective shows that a gene difference can define a new relationship with the environment. In this case the slightly unusual environment is phenylalanine in the diet, and the interaction of this with the PAH gene difference (PAH is the gene that is different in PKU) causes the catastrophic effects of untreated PKU. What we need here is information about at least one side of the equation before we can understand the other. So, knowing about differences in PAH enables us to define these individuals as a subpopulation, which allows us to observe that although phenylalanine is necessary to the bulk of the population it is actually an environmental poison to this small subpopulation.

The simplest way of summing this up is to argue that if we know the common gene differences in the human population, then environmental aspects are more easily analysed. This relationship is now recognized as central to human health and a huge experiment is at present being proposed and designed in the UK: 500 000 volunteers over 40 years old or so (this is roughly the age when diseases and death start becoming common) will be asked to join a study that will seek to record details of their medical histories. Ultimately, when the technology allows, most or all of these individuals will also be tested for at least the common gene differences they have. These records will provide a huge amount of information about the relationship of gene differences to disease and also open the possibility of studying as many aspects of lifestyle as the volunteers are willing to divulge. At the time of writing (2001), it is unclear if the proposal will be allowed to proceed—ethical problems centred on the protection of privacy and individual dignity are key difficulties.

If the study does proceed then it will in time have enormous impact on health services, individual lives, and the lives of our children, because it will open up the potential to avoid diseases by tailoring our individual environment to best achieve this goal. It will force medicine to focus not on treatment, as is the case now, but on prevention. Will this be the end of medicine?

I have discussed almost exclusively events that are triggered by differences within genes or the environment, but the human body is not simply a machine that comes into being fully mature and formed. Quite the contrary, we are all the products of the intricately controlled processes of development, within which cells have a central role. Exploiting the capabilities of cells is something gene engineers partly do but is much more the activity of cell engineers— scientists who research into new ways the body might be repaired.

■ Cell engineers

Scientists are researching various methods to extend the rather limited potential of engineering genes; the repair of damage to tissues caused by accidents, age, or diseases is certainly one of the most interesting. Here cell engineers, not gene engineers, are important. It should by now be fairly obvious that we have only a limited idea of which genes control the development of which cells and tissues, and so it is unsurprising that gene manipulation is not yet part of tissue engineering. Quite recently, though, tissue engineering has taken an explosive leap forward as researchers have realized that merging two areas of study—the cloning of animals and of stem cells—has great potential for future developments in gene engineering. What I would like to do now, therefore, is discuss these two areas, starting with Dolly the sheep, the first mammal to be created as a clone.

Cloning ourselves—copies on demand

Clones are not unusual; we have met them before as identical twins but a huge amount of publicity attended the announcement of the creation of Dolly the sheep. What exactly was all the fuss about? A cloned animal is simply an animal that contains DNA identical to another, as two identical twins are identical. So Dolly was a clone because she was genetically identical to her 'mother' and had all the same gene differences as her mother: this is a unique relationship because identical twins are always related to each other and never to their mother.

Dolly's relationship to her mother comes about because she was made by a most unusual route: an udder cell was taken from Dolly's 'mother' and the region that contains all of the DNA, the nucleus, was carefully removed with fine needles and tubes and injected into an egg, from which its own nucleus had been removed. This process sounds technically difficult, but was routine to specialist researchers who have carried out similar experiments many times before. The experimental egg was then placed in the womb of a sheep (not Dolly's 'mother'), who had been injected with various proteins (hormones) to prepare her body and womb for pregnancy. The egg divided and developed and Dolly was born as a normal lamb some months later. Dolly, subsequent analysis showed, had the DNA and genes of her 'mother', who was the sheep whose cells were used to provide the nucleus, and so Dolly was a 'clone' of her 'mother'. There were over 270 attempts to make Dolly and all failed except for Dolly and two other lambs, both of whom died shortly after birth. This was rather a high failure rate, prompting the question as to why the process was so hard to do.

Dolly was not the first cloned animal—that honour falls to a frog many years ago—but she was the first cloned mammal and the intervening period had seen many attempts that were very similar to those that finally succeeded with Dolly. The main reason they had failed, and the only new approach used in making

Dolly, was that the cell from the 'mother's udder was in a particular state—it was not growing. Most cells in our bodies are actually in the same state—it is called 'quiescence'; it seems that the egg can do something very remarkable to cells in the quiescent state. When I discussed human development in Chapter 2, I focused on the way a cell becomes in part programmed by what its genes have done in the past and this is a vital part of the control of development; it is as true for udder cells as it is for cells in a developing embryo.

For an udder cell to become an embryo, all its genes will have to be 'reprogrammed' and there seems to be something in the quiescent state that allows this to be achieved by the egg. In contrast, genes in the nucleus of dividing, and therefore non-quiescent, cells cannot be reprogrammed. To be frank, we have not the faintest idea as to what is really going on in the reprogramming process. Not only do we not know which genes are becoming reprogrammed and how, we do not even know what a 'program' is. All we do know is that it works and it works rather inefficiently in sheep. Once it was realized that using quiescent cells was the key, successful cloning of mice, cattle, and pigs was rapidly achieved. The cloned mice were interesting because they were much easier to make, with a much higher success rate than for Dolly.

Why is all of this so exciting? Well in part because if you can do it with sheep, mice, cattle, and pigs you might be able to clone humans. This has led some scientists—some with absolutely no previous record of working on anything to do with this area—to announce that they would do just that and clone a human being. Nobody takes these announcements seriously because the technical difficulties would be extraordinarily high. Where are they going to get all the women to volunteer to lend their wombs to carry the baby, and what about the consequence of achieving success in only one out of 270 or so attempts? Who would volunteer to donate all those human eggs? Personally I do not take these claims as anything other than mild publicity-seeking. I have no doubts that the experiments could be attempted on a human beings but I have enormous reservations about the moral and ethical difficulties of the human 'logistics' of attempting them.

Unfortunately (or fortunately, depending on your views of the morality) there is a possible way round some of the difficulties of obtaining human eggs because the egg in cloning is just a clever way of reprogramming all the genes in the quiescent, donor, nucleus. The egg has no nucleus and so has no genes from the original egg nucleus. So if this is the case, why not use the egg of another animal? Why not use a monkey or a cow egg? In principle, after it has divided a few times, most of its proteins will be human because it has only human genes and so can make only human proteins. So very rapidly the embryo will be human, even if it started out as cow.

There have already been reports of attempts to undertake this experiment, with the egg allowed to grow in a test tube. The scientists claimed that the 'embryo' that they had created could have grown if placed in a womb, but other observers have commented that there have been many instances where an 'embryo' in a test tube appeared normal but was incapable of further growth, even in a womb. The experiment, which to my knowledge has never been published in a scientific journal, does not in any way prove the feasibility of this approach. At present, therefore, we have no idea if such a hybrid egg would indeed grow in a human womb and make a normal human baby. But the dilemma of the failed human experiment makes this a dangerously unprincipled undertaking. It is an experiment that could and, I guess, will be done in mice or sheep.

The moral implications of human cloning have been endlessly debated by ethicists in discussions that are all too often more firmly rooted in science fiction than the reality, which is that the practical possibilities for human cloning will remain very distant until there is some simple way of avoiding having to use a human being as the egg donor and as the incubator for the clone—a possibility that is firmly in the 'possible but unlikely' or 'impossible' category.

I want to step away from this speculation and look at an area that is scientifically realizable. This is the idea of growing cells in a test tube—so-called 'stem cells'—that have the potential to make a human embryo. These embryo cell cultures are more exciting to medical researchers than any possibility of cloning humans because of what they are and what they can do, and this is what I will discuss next.

Embryo stem cells—endless potential

Human embryos at 3 or 4 days old are little hollow balls of cells 0.1–0.2 millimetres across and in one area is a group of cells destined to become the embryo. These cells, very unimaginatively called inner cell-mass cells, are quite remarkable: experiments on mice have shown that if they are carefully removed from the embryo they can be grown virtually indefinitely in a test tube and, even after this, provided they are put into a womb, they retain the ability to turn back into an embryo.

The most important point to realize about inner cell-mass cells is that they have the potential to develop into every cell type in an embryo and, in turn, into all cell types of an adult. In development, a cell that produces many different types of new cells with new properties is called a 'stem' cell. Because stem cells in an embryo can give rise to a whole embryo, they are known as 'embryo stem cells'; these are becoming extremely important in research.

The commonest use of embryo stem cells is for making transgenic animals (mice in particular). It is easy to introduce DNA into the cells and, because they can subsequently be grown in a test tube, the experimenter can easily analyse them and make sure that they have taken up the new DNA correctly before they are placed in the womb to develop into a mouse—much quicker and cheaper than breeding mice to see if the new DNA is positioned correctly or is working properly. The work that went into making transgenic mice in this way is a fantastic scientific achievement and, without it, we would know very little indeed about the genes that control development. But this monumental achievement is not, perhaps, what embryo stem cells are going to be most remembered for; this is because these cells might provide the ultimate cure for a whole host of human diseases.

Physician, clone me—instant identical twins

Embryo stem cells have been made from human embryos—or we think so. The caveat is that the ultimate test for an embryo stem cell is that it must be able to be grown into a normal human embryo in just the same way as is done in mice: for humans this would obviously be a repugnant experiment (and now illegal in some countries). But all of the evidence up to this final point suggests that inner cell-mass cells from humans will, in a variety of experimental conditions, behave exactly as we would expect embryo stem cells to behave. Most importantly, the human cells seem to behave just as mouse embryo stem cells do when they are put into similar experimental test systems. The general opinion, therefore, is that even without the ultimate test of making humans, these cells are true embryo stem cells.

Why might the discovery of human embryo stem cells be so momentous? Simply because these human cells can easily be grown in a laboratory, frozen, re-grown, and stored for many years, and still have the potential to develop into every cell type in a human being: nerve, brain, eye, heart, egg, sperm, muscle—everything. If experimenters can learn how to control them, to learn how to make them develop into a given cell type on demand, then this opens up a wonderful future where, when new tissue is needed, all the researchers need do is to go to the freezer, unfreeze a vial of embryo stem cells, and make them grow and become new tissue. It may be the way that broken nerves can be rejoined and so to heal the paralysed; it may be the way we can halt or reverse the ravages of Parkinson's or Alzheimer's disease, which are caused by the death of brain cells; it may be the way we can grow muscle or bone to restore injured limbs. The list of possibilities is very long and reflects much human suffering.

Before this list of potentials and possibilities carries us away, I think it important to emphasize that there is a huge amount still to learn if the promise of these embryo stem cells is ever to be translated into reality. Medical scientists do not yet know how to make them grow into particular cell types; although something is known about this, nerve cells or muscle cells cannot yet be made to order. The next big problem is that stem cells turn into embryos only when they are growing in a womb. It's not clear yet whether every cell type can be made in a test tube nor whether whole tissues rather than just unorganized cells will grow in this way; and there is much uncertainty about the myriad controls over cells that are exercised during the complicated period of development in the womb. Finally, and perhaps most critically of all, we should not forget that these are embryo stem cells, not adult stem cells; they make all of the cells of an embryo but these are not always the same as cells in an adult. There are many cases where embryos have cells with properties that are different from those of the adult—for example, in humans, red blood cells in embryos are rather different from those in the adult—and so there is no guarantee that, say, a nerve cell from an embryo could replace a nerve cell in the adult.

Nevertheless, there is huge excitement over the potential for therapy using stem cells and this excitement grew when it was realized that a combination of human cloning and stem cells could have a quite remarkable experimental result. It is important to understand that if I was to take a piece of skin from my hand and surgically implant it into another human being, the skin tissue would rapidly be recognized as 'non-self' (see Chapter 5) and would be rapidly destroyed by defence cells. This is why all forms of transplants—heart, kidney, liver, bone marrow, skin, and so on—are so difficult, because the defence response has to be suppressed with powerful drugs. The ideal in conventional transplantation is that you choose a donor with MHC proteins whose gene differences are as exact a match to the recipient as possible, thus fooling the recipient into thinking the donated tissue is actually 'self'. The best case is to use the tissue from an identical twin so both donor and recipient will have exactly the same gene differences, but identical twins are comparatively rare—until now.

Now, perhaps, everybody can have the equivalent of an identical twin. This remarkable feat could be achieved by first making a clone of yourself and then making embryo stem cells from the clone: because these would be derived from your own body, they would be a perfect source of 'self' cells. Medical researchers recently carried out this technical feat in mice. They took quiescent cells from an adult mouse and used these as the start of a cloning experiment; the resulting embryo was not put into a womb but was instead allowed to grow in a culture medium for about 4–10 days. At this stage of development, the embryo has an inner cell mass and the cells were picked out and grown separately as embryo stem cells but with the unique property of being identical—down to every last gene difference—to the cells of the original 'mother' mouse. This is remarkable. If this process could be reproduced in humans (and every step is conceptually or practically possible), then not only could you have a cell resource for potentially making many different tissue types in your body, you could also guarantee that every embryo stem cell was genetically identical to your own cells: you would be in the position of having your very own identical twin!

This process is bound to become simpler with time; once we have a well-characterized human embryo stem cell it will be reasonably simple to tailor this to an individual's gene differences by simply removing the existing nucleus in the cell and replacing it with one taken from whoever happens to need the cells.

Tissue and cell engineers have a huge amount of work in front of them with some exciting and challenging goals. Are embryo stem cells as exciting for the gene engineer? Perhaps the most exciting contribution might be to provide a potential route to a controlled gene therapy, where the new DNA can be carefully studied to check for potential gene damage at its site of insertion before the cells are reintroduced into the patient. Certainly such an approach might avoid some of the dangers of conventional approaches in adult and embryo gene therapy.

The role of embryo stem cells in gene therapy is speculative and may ultimately not be important. But for the gene engineer, using human cells to study development is a virtual certainty: the cells can develop in a test tube, even if they will not go so far as to make a full embryo, and this will open extremely powerful experimental approaches. Genes that control major developmental stages in humans could be studied from their effects on the development of stem cells rather than human embryos.

Perhaps this is how cell engineers will learn to make stem cells develop into particular tissue types—by deliberately turning on and off the genes that naturally control the process in the more complicated environment of the developing embryo. Ultimately, but we really do not know what that means in actual time, it may be possible to trigger some parts of development in a test tube and perhaps make whole organs—for example, a new liver to replace a cancerous one, a new eye to replace a damaged one, a new heart to replace the old; the potential is vast. So let's turn to this area and see what might be possible in the way of making whole organs, surely the ultimate achievement of cell engineers?

■ The body shop

Organ transplants, a dream just 40 years ago, are now routine and an enormously inefficient way of meeting the need for replacement parts. Quite simply and brutally, not enough young people are getting killed so that their organs can be used for transplantation, and there are far more people waiting than there are organs to be used. The difficulties of obtaining a good tissue match simply exacerbate an already complicated situation. For these reasons, the potential for growing organs is enormous and it is fairly clear that there could be some unusual routes to doing this that are firmly based on the reality of our understanding of the genes that control development. These approaches are not only unusual, they also push at the limits of what might be thought to be allowable in human beings—indeed, to the very roots of the definition of a human.

Headless babies—spare parts to order

Perhaps I'll start this discussion by asking you to think about something that is probably disturbing and distressing: what would you do with a headless baby? Unfortunately the birth of such a poor creature is not so rare, perhaps one in 10 000 births. The babies lack the main parts of the brain—the fore- and midbrain, rather than the whole of their heads—but in every other respect seem to be well formed and the rest of their bodies are unaffected. These poor babies can survive hours, days, or, very occasionally, weeks. Given the tremendous shortage of suitable donors, it is fairly unsurprising that such babies should be used as donors for transplants of various organs and this has prompted some very difficult thought within the medical community. Foremost was the consideration of whether such babies are really human: the general, probably universal, agreement was that they are, despite lacking any ability to think or respond as a human would.

The question that until now has not been really answered is exactly how much of a human must be left for us to continue to consider 'it' human? At one end of the spectrum, no one could reasonably argue that a one-legged person is any less a human because he or she lacks a leg, but what of one leg that lacks a body? A very strange idea, but how else would you describe an embryo that consist only of leg and not any other parts that we consider human; would this be human or not human?

Many people might find this line of thinking repugnant and I do not pretend that it is a simple matter of either science or ethics, but it is a future that is in all probability approaching; it could well be an important way of preventing much human suffering by enabling us to 'make' new organs to save lives that otherwise would be lost. This is surely worth thinking about.

I think it might be helpful to call these rather freakish fragments of human embryos 'embryoids' to distinguish them from viable embryos. Is it possible to make embryoids? We can avoid some of the moral dimensions of this question by looking at experiments in mice, where several gene regulators are known to be required to make the head tissues. The clearest example is the protein called Lim1, which is a distant relative of some of the Hox genes that I discussed in Chapter 2. A transgenic mouse is made that lacks both Lim1 genes; it is born lacking virtually all of its brain, an abnormality that can be caused by the lack of several other proteins.

As the explosive growth of knowledge about the genes and proteins involved in controlling development continues, we will certainly understand more and more not only about normal development but, inevitably, about how development can go wrong when proteins are missing. This is an inevitable consequence because eliminating genes by transgenesis is a powerful way of studying their normal role. It seems likely that research on how a mouse makes a leg, for example, will result in mice born with legs missing and also mice born with too many legs. In the future, there will be increasingly large numbers of ways of making embryos that lack critical parts, so it will probably become possible to make a mouse with no head, no stomach region but just a chest and a heart, or has rear legs but no front legs or head.

In principle, such mice do not even have to be made by transgenesis because already there is some evidence that approaches similar to those used in gene therapy can be aimed at 'blocking' the normal working of a cell's gene. The way this can be done is to use a virus that carries the gene you wish to block, much like normal approaches to gene therapy, but, this time, the gene is back to front in the virus—it is an 'antigene', if you will. The antigene makes RNA, but instead of making messenger RNA, as the normal gene does (if you need refreshing see Chapter 2 for all you need to know), the antigene makes an 'antimessenger'

RNA that is a perfectly complementary match, rung for rung with messenger RNA and so the two can make an RNA version of the double helix of DNA. Messenger RNA has to be one-stranded to be used in the cell and so the effect of making it look like the DNA ladder is to prevent any protein being made from it. In essence, the presence of the 'antigene' has resulted in loss of the protein encoded in the gene. This approach is not yet fully effective but, nevertheless, in many cases genes can be switched off in this way and one can see how it may then be possible to tailor how an embryo grows by infection with different viruses carrying different antigenes.

From the point of view of our discussion, it is clear that mouse embryos can already be made that could never survive as mice but could readily be used for spare parts. If we take this knowledge, and combine it with the ability to make and manipulate stem cells, then the time of the human 'embryo shop' is not so far off. We should now be developing the framework to control what could be done in the future because, if we wait, we will again be condemned to trying to catch up with a technology that is developing faster than our control of it.

Perhaps the most powerful argument against such technology is that you would need to use a woman's womb to allow the embryoid to develop and this idea, to me, is repugnant and morally unsupportable—a woman cannot be considered as a convenient womb. Of course, if you decide these embryos are not human, then there seems no reason why the womb necessarily has to be that of human; why not the womb of another species?

Much of what I have written will be distasteful to many readers: indeed, much of it is distasteful to me, but it is only by exploring what medical genetics might achieve, as opposed to what it has achieved, that we can start to consider the future and how we might control and benefit from gene research. The simple option is always to ban what we are frightened of, but to do so would be an appalling decision because about 2 per cent of all children born in the industrialized world suffer from some form of developmental problem (the figure for children in the developing world is unfortunately unknown; it may well be greater). It is only by understanding the normal process of development that we will ever approach the possibility of being able to help these children. Such research might even open up the last great dream of the genetics of development—to develop the ability to make the normal human body regrow and regenerate itself, because if we can do this we will need no embryoids, no wombs, no stem cells; we just need the natural growth of the body. It is to this final possibility that I now turn.

Growing a new leg

If only we could grow a new leg when the old one breaks! This must be the ultimate dream of medical science but it is also a dream that to be realized must be based on a detailed understanding of human development. Many living creatures have remarkable powers of regeneration: plants can be regrown from just single cell and starfish can make a new five-armed body from just a small fragment of one of their arms. This is a truly remarkable achievement; imagine that a human leg could regrow the rest of the human body and you will see why. The most remarkable animal known so far is probably a little creature called *Botrylloides*, a sea squirt, that lives on rocks in shallow parts of the Mediterranean Sea. Researchers have been able to regrow a whole organism from a tiny part of the blood system—cloning made very easy indeed. In contrast, the best that we poor mammals can do is to regrow parts of an organ that have been cut off, provide not too much has been removed.

The sort of regrowth we are interested in needs cell division, which surprisingly does not have to be a feature of regrowth in animals such as the simple *Hydra*: cut a *Hydra* in half and it will form two smaller animals with half the number of cells of the original. Most animals have to make new cells to regrow an organ. Perhaps the most impressive achievement of this type of regrowth is that of the newt that can regrow, amongst other tissues, whole limbs, its jaw and tail, and even parts of its eye. We are far from understanding how the newt achieves this feat, but many of the proteins involved are the same set of signalling proteins and partners as are active in normal development, and the crucial events happen very early in the process of healing. For example, after a newt loses a leg, cells from the skin rapidly grow over the amputated stump and these seem to be able to trigger the cells below them to reverse their development and become 'reprogrammed' to make a new leg. This is a very unusual ability for a cell to have because in most creatures, including humans, development is irreversible; in a sense, once a cell has become a leg cell, it can never become anything else.

There is also a good deal of evidence that relatively simple chemicals can have a profound effect on regrowth. The best example is a chemical called retinoic acid, a very close relative of vitamin A. If a newt's hand is cut off, a new hand will develop, but in the presence of retinoic acid in quite high levels a whole new limb will grow back onto the handless limb. Equally bizarrely, if the tail is cut off a frog tadpole, the presence of retinoic acid can make a regenerating tail grow limbs instead of a new tail. Retinoic acid (or perhaps a close chemical relative) is a partner compound to a signalling protein and it is likely that its amount varies along the body's axes in a gradient that cells appear to use to define their position relative to the axes: the aberrantly high concentrations signal the wrong positions and so trigger a response that is inappropriate to the cell's true position.

The key challenge is whether this knowledge can be used to trigger redevelopment of human limbs. Much is now known about this sort of regeneration and what proteins and chemicals are important, but the fact remains that scientists are totally at loss to explain why it can happen in newts and frogs but not in humans. Our greatest ability in limb regrowth is for children to regrow the tip of their finger provided the cut is not lower than the nail. One possibility is that the complexity of making a mammal is simply much greater than for a newt, even though both have fundamentally the same body plan. Perhaps developmental steps are so complicated that they cannot be reversed? We simply do not know.

One piece of knowledge is tantalizing: to clone an animal such as a mouse the egg must be able to 'reprogram' the genes in a quiescent nucleus. How does it do this? If we knew, perhaps we could use this knowledge to reprogram cells in the adult, such that they might regrow an arm, much as cells in the embryo originally grew the arm. At present, however, nobody has the slightest idea how to reprogram or reinitiate organ growth. Even if we did know, all the complicated three-dimensional interactions of cells touching other cells and of signalling proteins and near and distant partners would probably have to be mimicked to achieve the final goal of a functional arm. At present this is far beyond the limits of our understanding: it is impossible practically. To sum up, then, until we have a better understanding of the processes of development and how these are controlled, and how our genes and environment interact, there is no chance of retriggering the growth of limbs and whole organs: this is one area where dreams will probably remain dreams for a good few years.

In conclusion

It is a something of paradox that non-scientists are most interested in those areas of genetics that scientists know least about. The lack of knowledge is a problem to both sides—to the non-scientist who erroneously sees geneticists as all-powerful figures and to the scientist who can never meet society's expectation. The truth is often far more prosaic than we might like and the ferocious biological complexity underlying our genetic inheritance means I can summarize the current state of play in this field of science as follows: 'we have a severely limited ability' to alter our inheritance or its effect.

There are several reasons why this should be so:

- Many genes contribute to the important features of humans that geneticists seek to alter; each gene difference has a tiny influence and so many genes would have to be altered.
- The interactions of gene differences and environment are in most cases hidden; changes to genes may be reversed, neutralized, or reinforced by our environment.
- The influence of gene differences may be exerted in development, and so the effects on adult bodies will remain to all practical purposes irreversible; changing some genes in the adult will therefore be ineffective.
- Given that a single gene cannot be altered at will, the prospect of altering many genes simultaneously is a fiction.

The technical limitations today, therefore, are substantial: we understand little about how DNA is introduced into cells and its fate when it enters, and we have an incomplete knowledge of the gene regulators that control any gene, and of the effect of cell history or cell surroundings. Until this ignorance is rectified, my fellow geneticists and I are proceeding by informed guesswork.

The summary of the future prospects is only a little different. I would add just three words to what I've said above: 'we will continue to have a severely limited ability' to alter our inheritance or its affect. This is not true across all fields; understanding of gene differences in cancer is becoming so profound that, in a relatively short time, I believe that the gene events that precipitate cancer will largely be known and so there may well be a simpler gene target than would be the case for other human gene disorders.

In contrast to this relatively pessimistic view of the future, which is, after all, simply an appreciation of the biological reality—and not a criticism of current research efforts—the prospects for reintroducing genes in the future can only improve. Already scientists are on the second, third, and even fourth generation of systems for introducing DNA into target cells and these are perhaps a thousand times more efficient than the first-generation methods. This progress cannot continue indefinitely. We are still no wiser about the control of events after the introduction of the DNA, but my guess is that there will continue to be improvement: its extent will simply have to be left up to the mass of researchers who are chipping away at the problem.

Given this view, I do not think there is any prospect for improvement in the complex abilities of our children, such as their intelligence, beauty, or strength: to think otherwise is science fiction, posing as science fact. We all have Genetic Destinies, and to look to gene therapy for our salvation is to seek to grasp a mirage. Nevertheless, geneticists will help remould individual destinies in our children's lifetimes, by understanding the interplay of gene differences and environment. We will be faced with wonderful new opportunities to cure. Our individual Genetic Destinies will become modifiable but perhaps not escapable, and the supreme challenge of genetics is to ensure that this power is used for the good.

Dreams and nightmares revisited

This book started with the future histories of Jeanne and Jean whose epitaphs were fundamentally different:

> *Jeanne Dream lived and died in the future, where disease and suffering were an echo of the past.*

> *Jean Battler lived and died in the future, where disease and suffering were the results of nature and malign human influence.*

We can now revisit these tales with a much firmer grasp of their underlying scientific principles and look at the reality behind the stories. In doing this, we have three goals in mind: the first is simply to decide what is possible, what is impossible, and what lies in between these extremes. Our second goal is to draw lessons as to why the lives of Jeanne Dream and Jean Battler are so incomparably different. By the end of this chapter we will have met each of these goals and we will be ideally positioned to confront the fears that surround much genetics today. To understand the source and reality of our widespread distrust of this science will become our third goal.

In the re-telling of the future histories that follow, I have simply reprinted the stories inside text boxes. Each story is identical to the account in Chapter 1, but now interspersed with unboxed text that seeks to discuss the context of each section of the life histories.

■ **The gene dream**

> This is the future history of the life and death of Jeanne Dream, who was born in the small town of Prosperous. Her parents, Perfect and Manley Dream, had been married for 5 years before moving to Prosperous to avoid the high taxes and crime of the City and to find a safe place to raise a family. One year before Jeanne's birth, they both attended Dr Goode's surgery, where DNA tests were run to check for unfortunate gene differences that might give rise to problems for their baby. They were also counselled as to the possible range of intelligence, hair colours, skin complexion, and major personality attributes their baby might possess.
>
> Heartened by this information, Perfect and Manley go home and, as in the best stories, they share a wonderful meal together. The next morning they return to Dr Goode for an overnight stay in her Love Ward. Several more visits follow, at the end of which Dr Goode has taken 10 eggs from Perfect and fertilized them with sperm from Manley in an *in vitro* fertilization (IVF) operation.

☐ **Possible.** The two key events we need to consider are, first, the screening for DNA differences to advise Perfect and Manley on the possible characteristics of their baby and, second, the use of IVF. Genetic screening is routine in many industrialized countries and the technology for identifying changes in DNA is rapidly improving; this is reality. Currently, however, gene therapists do not know the identity of enough DNA differences or how these changes affect us to be able to make useful forecasts with any accuracy. All of the predictions are possible but some may ultimately turn out to be impossible practically, because of the combination of many gene differences with minor effect and the complex confounding interactions with the environment. Only research will resolve the extent to which prediction will be possible.

At present, the best gene therapists could do would be to screen the parents to make sure they were not carrying one or other of the 100 or so gene differences that cause the rare, or fairly rare genetic conditions. If any gene differences with possible adverse effects on their unborn child were identified in Perfect and Manley, they could be advised on the chances of such gene differences being passed on and what would be the likely effects on their child's life. This advice would be couched in terms of probability—10 per cent, 20 per cent, and so on, and not in terms of certainties. Perfect and Manley could also be offered the chance of having their baby tested at about 4–6 weeks of pregnancy by having a small tissue sample removed with a needle and subsequent DNA analysis. In the event of the baby carrying a serious genetic defect, Perfect could be offered an abortion.

The second point is reality: IVF is routine and there is no reason why Perfect and Manley could not expect, if not 10, then certainly more than three embryos.

The fertilized eggs develop normally and at 10 days cells are removed for standard DNA testing and for possible banking of embryo stem cells. The results of these tests are presented to Perfect and Manley.

Embryo 10 looks particularly promising: it is female with a very low chance of suffering from any of the disorders caused by single genes. There is a 20 per cent of common cancers occurring; Alzheimer's has a 50 per cent chance, major depressive illness 10 per cent, and schizophrenia 1 per cent. The future child's personality profile is extraversion 60 per cent, neuroticism 10 per cent, agreeableness 70 per cent, conscientiousness 70 per cent, and culture 60 per cent. The only bad news is that the possibility of dyslexia, addictive personality trait, and attention-deficit disorder is 90 per cent, and there is a potential for criminal actions. But the gene counsellors advise this is 99.9 per cent unlikely to occur within the socially advantageous environment that embryo 10 will occupy. She will be red-haired, pale-skinned, and freckled—and be at risk from skin cancer.

Delighted by most of this, Perfect and Manley choose embryo 10 for stem-cell culture, and this is implanted into Perfect. The pregnancy proceeds normally; 9 months later, on 31 August 2020, little Jeanne emerges into the world after an uneventful pregnancy. Jeanne is immediately retyped for the same set of disorders to confirm the absence of DNA changes during development; all the results replicate the earlier ones.

☐ **Possible and possible but unlikely.** All of this passage is possible, but today, as I have already mentioned, geneticists do not know the gene differences involved in any of these conditions nor how the environment is conditioning their effect. Using DNA differences routinely to predict criminal activity falls into the 'possible but unlikely' category; at present, there is no evidence that gene differences indeed contribute to anything as nebulously defined as 'criminality' and the rare cases such as MAO deficiency (Chapter 6) have so far been found only in one extended family. Environmental influences remain the dominant contributor to criminal behaviour but the impact of behavioural genetic research in this area is likely to provide new insights into our ideas about the basis of antisocial behaviour; this area is likely to be a minefield in the coming years.

The complex relationship between genes and the environment is implicit in the gene counsellor's underplaying of the chance of a criminal outcome. The assumption is that Perfect and Manley will be able to provide a supportive environment for Jeanne's childhood and that this will be less likely to lead to criminal behaviour than one that was less optimal. This assumption may not be correct because we do not know in detail what environment predisposes to criminal behaviour and we must also recognize that environment contributes to an increased chance, but not a certainty, of an outcome. No gene counsellor could state that, because of her environment, Jeanne will not be criminal: all that could perhaps be stated is that it is less likely that Jeanne will have criminal tendencies.

Retyping Jeanne's DNA is a wise move to avoid the possibility of a new gene difference being present as a result of a DNA copying error made during the production of the sperm and egg that together gave rise to Jeanne. These types of changes would, of course, not be detectable by analysing DNA samples taken from either Perfect or Manley.

> She is also DNA-tested to detect her immune-system MHC (major histo-compatibility complex) type. Based on these results she is immunized against TB, measles, diphtheria, and five common cancers using vaccines profiled to her MHC gene differences. She is shown to be HIV-resistant and so is not immunized against that disease.

☐ **Possible.** How MHC affects the vaccination response in individuals is unknown. The problem is once again one of recognition of 'non-self'. Some differences within the MHC means that the individual who contains them is less capable of recognizing molecules, often proteins, in the vaccination as being non-self and so it is less effective in producing immunity. Tailor-making vaccinations to an individual's MHC gene differences is possible but a more likely immediate development is to use this type of information to tailor vaccines to larger groups who share changes, rather than basing them on an individual-by-individual basis.

The idea of immunizing against some cancers is starting to be reality; for example, the human papilloma virus (HPV) that commonly causes genital warts appears to have a significant role in later causing the development of cancer of the cervix. Trials are in progress to look at the efficiency of a vaccination against HPV. But such examples are uncommon and in most cases vaccination is not an option.

HIV resistance is in principle easily typed by looking for the DNA differences in the CCR5 signalling protein but, as I discussed in Chapter 5, HIV can use other signalling proteins to enter a cell. This, in turn, means that infection is possible even in people with the CCR5 gene difference and, as yet, we cannot predict that a person is completely resistant to HIV.

> Finally, Jeanne's DNA is tested for gene differences to the 155 genes that encode the proteins on the 90 per cent drug-response indicators list compiled by GlaxoSmithMerkNovartis—the GSMN-90—to ensure any future drug treatment is matched to her particular profile of variation. This information is recorded on a remote readable chip that is implanted under the skin at the back of her neck.

☐ **Possible.** We do not know how many genes modulate our responses to drugs: I mentioned one example in Chapter 5 in relation to the drug debriso-quin and the cytochrome P-450IID6 gene difference. The cytochrome protein was involved in breaking down the drug and about a dozen proteins with similar actions are known; any differences within these either already has, or will soon be, identified. It is unclear exactly how many proteins are involved in these processes but we can be confident that most of the key ones will be identified soon. Variation in drug response, however, can be caused either by variation in how the drug is broken down or by variation in the target of the drug. Identifying target variation is a large job—there are as many targets as there are types of drug—and research is at an early stage. But within 5 years the commonest changes in the most important targets are likely to be identified. Additionally, responses can be made more complex because of the potential effects of the environment—in this case other drugs or other chemicals that we ingest in our food every day. These 'natural' chemicals themselves have to be broken down by many of the same proteins that break down drugs and so there can be a relationship between diet and drug breakdown that can make these studies more difficult than they might otherwise be.

Even with these reservations, the 'GSMN 90' is not far away. The potential for treating people with individually selected drugs is enormous and would replace the sometimes lethal consequences of the 'suck it and see' approach that is presently used.

The idea of coding this information on a chip is technically possible: the amount of information is tiny—just a gene identifier, a rung number, and a change identity; we already use chips to identify our pets.

The pharmaceutical giant GlaxoSmithMerkNovartis is a very plausible entity: the cost of developing drugs is absolutely enormous—at present about US$900 million to get a new drug to the doctor's surgery. This has meant that only the very big companies can afford the enormous costs of successes (or, conversely, cover the costs of failure) and the logic of the economies of scale and market share has already spurred several rounds of mergers. These giants are so big, however, that they are slow to respond to changes in science and much of the progress in using genetics in the discovery of new drugs is coming from relatively tiny biotechnology companies that can respond rapidly to the extraordinary pace of discovery: being small, they often have a precarious financial stability and can, and do, easily go bust, so there is a tough trade off here.

> When Jeanne is aged two, the barbecue she is standing next to flares up and she receives burns to 45 per cent of her right arm. She is rushed to Prosperous Hospital and treated successfully with a skin-cell culture from her bank of embryo stem cells.

☐ **Perhaps possible and soon.** My guess is that 'personalized' skin grafts might be one of the most immediate results of the embryo stem-cell culture research simply because skin cells are relatively easy to grow in test tubes. Right now, human embryo stem-cell research is in the earliest of stages, but expanding rapidly: watch this area!

> Four years old, and based on her new school's recommendation, she starts a 1-year course of gene therapy against attention-deficit disorder. This is followed by dyslexia therapy using the GSMN cognitive enhancer (Iqzac) and intensive teaching therapy. The outcome is positive, but the school reports that Jeanne is very shy and so she is treated with GSMN's latest social enhancer (Zocialac). The school is delighted at the results and the rest of her schooling is uneventful: she graduates with excellent grades and enrols at Prosperous Private University.

□ **Possible and unlikely.** None of this section is currently reality. Several studies have suggested that attention-deficit disorder may be in part due to change in a gene called DAT1 that is involved in transport of one of the important chemicals in the brain called dopamine, but these results are not yet conclusive. Consequently the idea of gene therapy is hopelessly premature: we neither know the identity of the differences within genes nor their contributions to the symptoms of the disorder—this is likely to be impossible practically.

Linkage analysis has suggested that at least three regions of human DNA must contain differences within genes that contribute to a person having dyslexia, but the work is far from complete; geneticists are still in the stage of attempting to replicate studies and, as far as I am aware, no actual genes have yet been identified.

Research into developing cognitive enhancers is underway in many laboratories but is severely hampered because we know so little about cognition, let alone its genetics and the effects of gene differences. Surprisingly, Zocialac is not quite as unlikely as we might have guessed; three studies, involving 800 people treated with a SmithKlineBeecham drug called Seroxat, have shown it is effective at treating a condition called social anxiety disorder: this affects about 2–5 per cent of people (commonly teenagers) in some US surveys, and the symptoms are an 'overwhelming fear of embarrassment' in simple social situations, such as group meetings, talking to the boss, or chatting over lunch. This leads them to avoid many of the day-to-day events that most of us face with equanimity. Seroxat is more normally used to treat depression, obsessive-compulsive and panic disorders—all complicated psychiatric conditions—and its use to modulate social interactions is unusual.

Just before she is due to go to university, on her eighteenth birthday, Jeanne is involved in a car crash and her left arm is badly crushed. She is left in a coma for 5 days. Rushed to the Prosperous GSMN trauma unit, her GSMN-90 chip is read and she is treated with an individually profiled selection of anti-inflammatory drugs to reduce the trauma to her head. Her crushed arm is treated with GSMN's Growzac, a bone-growth enhancer, which was discovered in 2010. Because of her relationship to GSMN, she is enrolled in a trial involving one of GSMN's latest products—'developmental restarting initiators'. She is treated with a cocktail of 10 naturally occurring developmental regulators that can reprogram the genes that are being expressed in the cells of a tissue so that they develop the capacity to renew themselves.

The treatment is successful for all parts of Jeanne's arm, but her hand remains irreversibly damaged and residual nerve damage requires further treatment with two compounds developed out of GSMN's 'developmental re-orientation of potential' (DROP) programme, which is seeking to develop ways of transforming of any tissue type into any other type, by reprogramming gene activity. In Jeanne's case, connective tissue cells are reprogrammed into nerve cells and the treatment of the damaged nerves is successful. Her hand, though, is in a bad way and her surgeons elect to carry out an organ transplant using a hand grown from an embryo by Embryofarm, a noted embryoid development company. The tissue match is perfect (it was selected to be so, based on matching DNA differences) and the organ grows perfectly, but Jeanne will forever have a slight mismatch of skin colour on her new hand. This is the subject of a legal case that is resolved after 2 years with Embryofarms admitting they failed to conform to their DNA difference standard for racial origin of 99 per cent purity.

Long before then, a graft from Jeanne's embryo stem-cell bank resolves the residual tissue loss on the scarred surface of her arm.

☐ **Possible and possible but unlikely.** There are four issues: personalized anti-inflammatory treatment, Growzac and bone regrowth, the DROP programme, and, finally, embryoids.

Inflammation is a symptom of the body's response to damage and infection and learning how to control it would have enormous impact on many different conditions. At present, there is a great deal known about the many signalling and partner proteins involved but we have, as usual, a far from a full picture and how individual variation and gene differences might influence our response is not understood at all.

Growzac does not exist but trying to understand and reverse the bone loss in a common condition called osteoporosis is a huge area for research. Osteoporosis is characterized by increasing bone fragility and is particularly common in postmenopausal women but may start as young as 25; in fact, all women have a 30–40 per cent and men a 13 per cent life-time chance of breaking a bone through the effects of osteoporosis. Treatments such as hormone replacement therapy are reasonably effective, but more effective treatments with reduced side-effects are certainly needed. The research into osteoporosis should identify the pathways that lead to bone loss and it is possible that this knowledge could equally be applied to understanding how to make bone regrow.

The DROP-related programmes are an invented name for all of the development research that is currently underway in the world; I discussed a little of this in Chapter 2. The critical question here is the potential of being able to reverse the cascade of signalling and partners and gene regulators: this may be a classic 'impossible practically' problem, where so many interventions are required that even if any one step is possible, the collective attempt becomes impossible. Equally, it is possible that stem-cell research will provide a short cut for such efforts, but at present we need to wait for the favourable results of a huge body of research.

Finally, embryofarms and embryoids: as I explained in Chapter 9, these are firmly in the 'possible but unlikely' category and will remain so whilst they require the use of surrogate mothers to produce the organ. Research on mouse embryo stem cells will power this type of approach.

> In an after-note to the case, the Embryofarm surrogate mother who incubated the hand, a young woman of 18 called Jean Battler, is fired for failing to disclose an accurate gene record. The company knows this is nonsense because they DNA-typed her themselves, but it enabled them to escape some of the legal liabilities.

☐ **All too possible!** The question that is being raised here is the legal status of the knowledge of an individual's gene differences, which is information that has a very different value to the individual or to an external agency. We are at the start of a long process of defining the legal status of such information and I will return to some of these issues in Chapter 11. For the present, we can be sure that once again the lawyers will not be out of work for a number of years yet.

> Jeanne starts university a year late because of her injuries and time in hospital. In her first year she shows the initial signs of alcohol abuse and is successfully treated with a combination of receptor blockers and serotonin analogues (GSMN's Banishac).

☐ **Possible.** Gene differences clearly contribute to alcohol abuse; different studies have produced results consistent with a 30–60 per cent contribution to alcoholism and related addictive personality disorders. There are indications that one gene difference must be within a particular region of human chromosome 4, but the identity is as yet unknown. Some studies have implicated differences near the DRD2 gene, which, like the DRD4 gene differences I discussed in Chapter 6, may alter dopamine levels in the part of the brain involved in our euphoria response. Differences within the DRD2 protein provide a logical way with which to explain excessive consumption of alcohol because they could mean that affected people develop a craving not so much for alcohol itself but for the effects on the body of the 'euphoria response' system. Unfortunately for this idea, 'logical' does not mean right, and several studies have contradicted the original findings. We are at a very early stage in this area and, as yet, there is no evidence for the involvement of either the dopamine or the serotonin system in this type of addiction. But if just one contributing gene difference was identified, it might lead to the identification of a pathway and open up enormous possibilities of developing a drug treatment.

> She travels to Africa for her middle year to help in an anti-tsetse fly eradication programme using tsetse flies that have been behaviourally modified genetically. Wading through a swamp to rescue her best friend, she acquires a bad infection of bilharzia, but this is successfully treated with GSMN's Billyhac that had been made after the identification of the drug target (a DNA-copying protein) in the bilharzia genome project in 2002.
>
> In 2041, Jeanne graduates with a good degree and goes to work in the pharmaceutical giant GlaxoSmithMerkNovartis as a financial analyst. She has a very successful career and social life for 15 years during which time she meets, and finally marries, a tall, wonderfully kind, blue-eyed, PhD-level GSMN gene scientist called Wilde Hope.

☐ **Reality.** Bilharzia is more properly known as schistosomiasis. The schistosome genome project is underway and the full rung sequence of the organism should soon be established. More importantly, similar projects have been initiated or completed on most of the organisms—bacteria in the main—that cause many of the common infectious diseases; at the last count, more than 70 projects have been completed or are in progress. The projects will identify all of the genes, and therefore the proteins, that are used by the organisms, and these in turn will become the targets for drug development in the future.

Within 2 years, Wilde and Jeanne decide to have children and Jeanne elects for a gene-therapy enhanced pregnancy because she is 38 years old.

Jeanne undergoes IVF, just as her mother did to conceive her, but on this occasion there are only three fertilized eggs for testing. The sex of all of them is determined as male and a gene scan of the embryos with the GSMN-99 DNA chip test for 99 per cent of all gene-related conditions, shows a less than 10 per cent chance of all conditions, except for four. The gene differences that predisposed to these conditions are rectified by hit-and-run gene therapy using GSMN's latest Hitandrunzac embryo gene-correction system. At Wilde's father's request, the baby's body size is augmented by 20 per cent and muscle bulk by 5 per cent, using GSMN's Bulkzac embryo gene-augmentation system. Nine months later Wilde junior is born and put down for sports education at a top private academy, Prosperous College.

☐ **Reality, possible, and impossible practically.** I have already discussed much of the testing for differences within genes—'gene scanning'—in previous panels. It is a mixture of reality and possible but, once again, it is very unclear how likely it will be that we can correct these differences by gene therapy. The 'hit-and-run' approach in this panel is a technique that works in a limited fashion and aims to alter the DNA to reverse a gene difference back to 'normal'. It is not, however, an efficient process and is best carried out on cells that can be analysed to identify that the correction has occurred. It would be, at present, impossible practically to attempt in a human.

Body-size augmentation therapy is impossible practically; at present we have no idea of which gene differences might contribute to extremes of size, let alone understand how they might be manipulated to make a larger person and this is likely to be the reality for a very long time.

Jeanne's post-natal depression is treated with Anxzac, a tenth-generation anti-depressant just released by GSMN and discovered as part of their behavioural gene programme that was initiated way back in 2005. She is DNA-tested at the same time using the new GSMN-99.9 chip. Her personal data chip, still implanted in her neck, is updated.

☐ **Possible.** Depression and its treatment is a major target of most drug companies' portfolio of research interests. The medical need is extraordinary because 12 per cent of women and 7 per cent of men will suffer from the disorder and twin studies suggest about 40 per cent of the variation between us may be accounted for by gene differences. At present there are no clear indications as to where the genes might be located, but research is made particularly difficult by the need to be sure that all the individuals in a study are suffering from the same forms of the condition: failure to ensure that all patients have the same disorder completely destroys the basis of any sort of genetic analysis.

Current treatments for depression can be reasonably effective; perhaps the most famous is Prozac, which modulates serotonin, the chemical 'messenger' in the brain that I discussed in Chapter 6. Several drugs that are partially successful at treating depression interfere with the proteins that control how much sero-tonin is present in our brains; this must imply that depression may in part be caused by changes in the depressed individual's control of these chemicals, even though there is very little understanding of how this contributes to depression. The success of current drugs quite rightly has encouraged more research on identifying other serotonin-modulating chemicals, but what is really needed is a better understanding of all of the pathways that contribute to the depressed state. This is where studies of gene differences will be important, but these are still at the very earliest of stages.

The GSMN-99.9 chip, of course, does not exist but the development of our understanding of the effects of gene differences will be continuous. We can expect the list of changes to lengthen with time, but we have no idea how many genes it will ultimately need to include or, indeed, if it could ever be possible to explain 99.9 per cent of all disease variation in humans by gene differences.

There are two good reasons why this figure might be unattainable. First, the extent to which DNA differences that are rare in the population account for an individual's disease risk and response is not yet understood. If rare differences are a significant factor, then we do not have the technology to identify these because it requires far too much analysis—ultimately a Human Genome Project on one individual. Second, we must not forget that our environment influences many disease susceptibilities far more than do gene differences.

Now 43, Jeanne is in the prime of life; 20 years of ecstatically happy life follow. This is punctuated by the occasional visit to Dr Goode's surgery for a variety of minor infections and ailments, all treated with programmed anti-bacterial and immune-response enhancers that are individually profile-matched to both her GSMN-99.9 chip and to GSMN's pathogen DNA-sequence database, which contains the gene structure of 90 per cent of all known human pathogens.

☐ **Possible but unclear.** Understanding an individual's DNA differences should, in principle, give us information about how the person will respond to infection, but how predictive of outcome this information will be much depends on how many gene differences are involved and what the environmental influences are. At present, with the exceptions of the genes I discussed in Chapter 5, we have no detailed idea of the contribution of gene differences to most diseases. All of the information suggests that the relationship will be complicated—more like the multiple gene differences in the case of the malaria-resistant mouse rather than the single Nramp gene difference in the mycobacteria-resistant one (Chapter 5). We know almost nothing of how specific changes in our environment might influence our susceptibility to diseases. If the effects are significant, as in all likelihood they will be, then the chances of being able to make useful decisions based on gene differences will be much reduced. The issue will need reconsidering in a few years time, once the genetic studies have been completed.

Wilde takes very lucrative retirement from GSMN, and in the same year Wilde Jr takes a silver medal at the 2085 international games (he is entered into the drug-free but gene-augmented category of the strongest fastest section).

☐ **Impossible.** Unsurprisingly, the chances of gene-based performance enhancement being reality are very slim; it would seem much more probable that we will see drug-assisted and drug-free categories in professional sport, long, long before we have to worry about gene augmentation. The athletes at an Olympics are prime examples of the importance of gene differences and environment: most of them have physiques that uniquely suit the sport in which they specialize, so, for example, the sprinters are heavily muscled and the distance runners wiry. Many of these characteristics have a strong contribution from gene differences.

Gene differences that may be relevant to athletic ability have been studied. In one case, a gene encoding a protein called ACE had changes that were more common than expected in individuals with high athletic ability. ACE is involved in the control of heart function and is a common target of drugs that treat heart disease, so it was an attractive idea that there may be a connection between the ACE gene change and the highly efficient heart function that is needed by most athletes. The study looked quite compelling until more detailed research was performed on the changes of the ACE gene within and between populations of people. This showed that the picture was more complex than the relatively simple one described in the original studies and the association looks much less clear cut.

Gene differences are not enough to win an Olympic Gold—this needs both gene differences and training. Without the physique, though, training will never be enough. The area where we need much more information about gene differences, and urgently, is in testing for drug taking. Drugs are detected in athletes by looking either directly for their presence in blood or urine or by looking for their breakdown products. We have very little idea of how different people vary in this respect and the reports in the papers of athletes who have failed drug tests but continued to protest their innocence (well, they would, wouldn't they?—as was said in a famous trial in the UK in the 1960s) might be a case of gene differences causing variation in the breakdown of allowable chemicals, to give non-permissible products in some cases. Equally, perhaps I am being too credulous.

We accept, indeed expect, that athletes are at the extremes of the spectrum of human abilities and it is unsurprising that we often know relatively less about these than we do about the majority of individuals falling well within the normal range. A particular example of this is testosterone, a partner protein naturally made by men. It has the natural function of inducing many of the male secondary sexual characteristics (muscle bulk, size, beards, and so forth). Testosterone is naturally made by women but in much smaller quantities than males, and supplementing this low level is a good way for women to increase their muscle bulk, if at the expense of becoming rather masculine; the drug testers are therefore very interested in the amount of testosterone in a female athlete. The problem is that some women, because of their gene differences, naturally produce high levels of testosterone; they are simply at the extreme end of the normal range and, in consequence, are likely to have rather masculine physiques, which may well give them a very substantial advantage in athletic competition against women with testosterone levels more within the middle of the range. Drug testers in athletic competitions must have a sufficient under-standing of the extremes so that these women can be distinguished from an athlete who is cheating by taking artificial testosterone; this is not an easy dis-tinction to make.

When Jeanne is aged 78, 15 years later, cognitive testing detects the first sign of Alzheimer's disease and she is treated with gene therapy to prevent the loss of brain tissue. Over the next 10 years she is treated once a year.

Two cancers are identified during these routine visits and both are treated with a combination of gene therapy aimed at forcing the tumours to differentiate into non-dividing tissues and immune-system stimulators that trick the body's defences into thinking the tumour is 'non-self', triggering it to destroy the cancer. All treatment uses GSMN's standard anti-tumour drugs and is 100 per cent successful.

By the end of this period, Jeanne is 88 and her anti-Alzheimer treatment is only moderately successful. She is placed on intensive cognitive-enhancer treatment (Iqzac, once again) but also on some third-generation DROP products to restore nerve-cell division.

☐ **Possible and possible but perhaps unlikely.** Jeanne has two medical problems here: Alzheimer's disease and cancer.

Her incipient Alzheimer's disease is a startlingly common occurrence; at 65, 10 per cent of people suffer and this increases to about 50 per cent at aged 85. The most distressing of the disease's symptoms are a progressive loss of memory and personality, which are caused by a progressive destruction of areas of the brain. The destruction is unusual in that it is accompanied by increasing numbers of unusual protein deposits called 'tangles' that contain, amongst others, a protein called beta amyloid. About 60 per cent of cases are of completely unknown origin but in the remaining 40 per cent some gene differences are more common than expected; in fact, these discoveries are a classic example—one of the few—of how linkage has been used to identify the gene differences contributing to a common disease.

One of the genes in an Alzheimer's sufferer, APP, encodes beta amyloid itself and is on chromosome 21—a striking discovery because children with Down's syndrome (Chapter 6) have an extra copy of chromosome 21 and so three copies of the APP gene. If you recall, almost all Down's children will go on to suffer from Alzheimer's disease at an unusually early age . The other genes where differences have been detected include those encoding the proteins APOE, PSEN1, PSEN2, A2M, and LRP1. Strikingly A2M can stick to LRP1, suggesting this is part of a pathway. But even this amount of information does not lead to a detailed understanding of how this dreadful disease develops or could be cured, so the idea of gene therapy is not yet realistic. There is also the problem that therapy cannot replace what has been lost: the death of brain cells causes loss of memory and replacement of cells would not necessarily replace the lost memories.

Of perhaps more immediate importance, the gene differences can be used for partially predicting the likely onset of Alzheimer's: if you do not have the gene difference called APOE4, the risk of developing Alzheimer's by age 85 ranges from 9 per cent to 20 per cent; if one copy of the gene is different, this rises to 25–60 per cent; and if two copies, the risk rises again to 50–90 per cent. Alzheimer's disease is presently both unpreventable and untreatable, and in these circumstances it is very unclear why anybody would choose to take such a test. The prospects for a therapy for Alzheimer's disease are much enhanced by these discoveries but are far from reality: nevertheless, this is very much a possible conjecture.

In contrast, both Iqzac and the DROP products to reinitiate cell division in nerves are probably many years away from development—possible but perhaps unlikely. In the latter case, we still lack an understanding of how and why the nerve cells have withdrawn from normal cell division. But the present progress in understanding the control of the cell cycle is dramatic and we can expect this research to result in at least an understanding of why nerve cells stop dividing, even though this knowledge does not have to lead to the ability to re-start them: we are in the hands of the biology of nerve cells.

Jeanne's cancer treatment is by a three-pronged attack: gene therapy, forcing differentiation of the cancer cells, and immune-system stimulation. All of these approaches are currently under investigation and all work to a limited extent. The gene-therapy approach is based on adding genes to cancer cells that produce protein that can then be targeted by drugs to kill the cell. Surprisingly not all cells need to take up the gene because there seems to be heavy 'bystander' killing (the dying cells seem to kill their neighbours). A major problem is 'targeting' the oncoming gene to the tumour and avoiding killing healthy, normal, cells—targeting is a problem common to all kinds of gene therapy. Overall, the approach is promising but far from producing a reliable cure.

The approach of tricking a cancer into differentiating is a very promising one: in essence, differentiation is another name for development and many cells, when they finally develop to their 'end' state, loose their capacity to divide. So, if one could make tumour cells differentiate or develop, they might stop the rapid and catastrophic division characteristic of most cancer cells. Current research shows that this can work in a limited number of cases. There are, however, relatively few simple ways, such as adding chemicals or proteins, to force a cell to differentiate and it is unclear how effective the approach will be: it is only likely to improve as our understanding of development improves.

The immune system is the technical term for the body's system of defence cells and proteins. There are many records of almost miraculous recovery from cancer and some of these contain reports of large tumours spontaneously disappearing; one possible way that this can happen is that the body suddenly recognizes the tumour as being 'non-self'. Why this should unexpectedly happen is unclear and it is a comparatively rare occurrence. The difficulty is that the proteins of cancer cells are composed of the same proteins as a normal cell and which the body has long recognized as 'self'; how this recognition is reversed is very difficult to understand. We need to know far more about our incredibly complex defence systems before any real understanding will emerge, and so this treatment is close to, but not quite, possible.

> Jeanne has noticed she can manage only two sets of tennis before her aches start to inhibit her movement; these arthritic conditions are treated at the same time using specific immune-system suppressants and bone regrowth programmers (again DROP products): success is complete.

☐ **Possible.** Arthritis is in part caused by the body's defence cells attacking the joints of their own body. Huge strides are being made in understanding how proteins control the defence response, but the immune process is so intricate that we may never be able to control it fully. A more reasonable prediction is that drug companies will develop increasingly powerful ways of modulating the defence response, which may eventually lead to new and more effective anti-arthritics. But whether we will ever be able truly to 'cure' the disease is open to doubt.

> Another 10 years pass with continual, successful, treatment for Alzheimer's and arthritis and the usual crop of minor ailments, including two or three cancers. Jeanne's heart is becoming somewhat enlarged but this is taken care of by a programme of cardiac de-differentiators that have been available for 30 years. During these investigations, a weakening of two blood vessels is repaired by a local application of cell-growth enhancers and cells from her embryo cell bank. As time progresses, she starts to suffer from some loss of muscle cells but this is easily treated with the DROP programme drug—third-generation ReBulkzac.

☐ **Impossible practically and possible?** There are several different treatments in this penultimate panel, most of which I have already discussed. The most noteworthy comment is the reduction of cancers to the 'minor ailment' category. Jeanne is 98 and living some 140 years in the future. I am pretty certain that many presently untreatable forms of cancer will be treatable by this time and this view is mainly based on the progress that has been made in the past 15 years of cancer research. We are now close to a unified understanding of why and how a cell becomes cancerous and this means that cancer treatment and research can start to focus on a selection of potential targets for drug development. This is a big breakthrough indeed—not one that made the papers but a breakthrough nevertheless.

Why was this story not in the papers? Because it was the result of many different researchers' patient efforts, which led to many interlinked discoveries rather than one single 'breakthrough'. To appreciate the progress requires an understanding of the whole field of cancer research—a huge and complex field of knowledge that is not easy for a single scientist to grasp, let alone a non-specialist (hence the lack of immediate press interest). In brief, cancer comes about because the cell accumulates gene differences that cause proteins to become non-functional or develop additional functions. These changes destroy the normal mechanisms that the cell uses to ensure its division is controlled correctly; there are several 'checkpoints' at which this happens. For example, proteins check to see that DNA has been replicated only once before a cell divides—twice or not at all is quite useless to cells that have to develop normally—or they check to ensure that an accurate copy of DNA has been made with limited rung copying errors. Once these checkpoints start to fail, one consequence is that the cell will accumulate more DNA differences, which in turn cause more systems to fail in a vicious cycle of errors and effects.

The triumph of recent cancer research is that we now know the identity of most of the key protein players—the 'gatekeepers' at the checkpoints—and so have a great new opportunity for designing strategies and chemicals to destroy the cancer cells. This will still not be easy because all dividing cells in the body, cancerous or not, have the same checkpoints and we face the difficult problem of forcing specificity on the treatments; for example, how do we kill the cancers and not other dividing cells? I predict, therefore, that cancer will never be a 'minor' ailment—it is impossible practically.

Jeanne's heart problems are much the same as many of us face today. An understanding of how the heart develops has progressed with extraordinary rapidity over the past decade and a start has been made on a complete picture; as more details emerge the reasons for the enlargement of the heart will also become apparent. I predict that the first drugs to combat this common and potential lethal condition will emerge over the next 5 years: this is science of the present and so very possible.

> Finally, on the last day of August 2140, Jeanne and Wilde lie down on their bed, together for the last time, and both swallow a single pill that contains a complex mixture of muscle relaxants, cognitive stimulators, neurotoxins, and nerve-function inhibitors. With music playing gently in the background, and hand in hand, Jeanne and Wilde gently and happily die, content in the knowledge that theirs has been a wonderful and fulfilling life, untouched by pain, diminution of abilities with age, and sure of the rightness of their decision to leave life together.

☐ **Reality?** The gentle death of Jeanne and Manley is not the product of gene science but it could be the product of other sorts of medical research. Perhaps death will be given the same importance as birth in medical research; there are, after all, only two consistencies in human existence that are truly universal—that we are born and that we will die.

> Jeanne lived and died in the future, where disease and suffering were an echo of the past.

☐ **The reality.** I think that all of us would like to live in a world free of the suffering and pain caused by disease but, of course, these are not the only causes of suffering. Jeanne lived and died in an imaginary world that was wealthy, that could carry out all of the scientific research required to make discoveries, and that was free of the impact of mankind's more destructive activities—war, oppression, and discrimination. Without such freedom, Jeanne's life would have been very different. How different? Let's now consider the gene nightmare of Jean Battler.

■ The gene nightmare

> This is the future history of the life and death of Jean Battler, who was born in the small town of Poorsville. Her mother was called Struglin' by everyone who knew her—she could no longer remember a time when anyone had

called her by her real name of Hope. Struglin' had no idea who her baby's father was—there had been several possibilities. But none of her boyfriends had stayed around long enough to find out she was pregnant, and she was not overly worried by this.

Struglin' gives birth to Jean in Poorsville district hospital, and under the public-health maintenance scheme, Jean is immediately DNA-tested for the usual set of major social-impact conditions. These show that she has a 50 per cent chance of developing Alzheimer's disease and an 80 per cent chance of breast or ovarian cancer. Struglin' is advised by her birth para-medic that Jean will never be eligible for the state medical insurance scheme. Because Struglin' has no money to cover this in any event, she really does not care. Jean's IQ is established by intelligence-gene testing as being of no greater than 80—well below the minimum 90 required for entry to the state education system.

☐ **Reality and possible.** There is no particular significance to Jean's country of birth—it could be in any country where gene research is a significant feature of the culture. In practice, this means most European countries, North and some parts of South America, Australasia, and some Asian countries.

Jean's first brush with genes is in her potential susceptibility to cancer of the ovaries and breast and to Alzheimer's disease. We have discussed both of these before, so we know that it is possible, even now, to obtain this type of infor-mation by analysis of differences within the Brca1 and Brca2 genes for ovary and breast cancer and within the APOE gene for Alzheimer's. In neither case would prediction approach the 80 or 95 per cent level. What is new here is the crude way the information is interpreted: an 80 or 95 per cent risk is being taken as though it means 100 per cent, whereas it does not. There's a chance that Jean will not suffer from either disease; we cannot know her individual fate, only that it is more likely she will suffer than someone without such gene differences.

Second, the information is being used by the state and the insurance com-panies to reduce the costs associated with her higher risk status. Because her gene differences are being treated as an inescapable facet of her existence, Jean is condemned without any prospect of escape. This is an intolerable load on her life. By what right has the state to determine the future of one of its citizens in this way? The insurance companies, of course, wish to eliminate Jean as being a higher risk than other 'normal' people, but, as I have argued at numerous points, all of us are 'abnormal' in some respects. The problem with both the state and the insurance companies is that they are taking a single feature in isolation whereas the reality is that about 80 per cent of us will die of a disorder that is influenced by genes and so, logically, 80 per cent of us should be refused insurance. This is, of course, an absurdity. Perhaps as our genetic understanding becomes greater, we will develop a more sophisticated attitude to the relationship of gene differences to individual life histories.

The ability to predict IQ from gene differences currently eludes us and is likely to continue to do so—I discussed this in Chapter 8. One clear point that I hope emerged from Chapter 8 was the relationship of IQ to actual performance in the world. IQ is probably a reasonable measure of collective ability—on average, a low IQ is probably a predictor of low outcome—but this does not mean that because Jean has a low IQ she will necessarily fail in life. Any society that allows the confusion of collective outcome with individual outcome is going down a very dangerous route, one that geneticists have to resist strenuously.

> More worryingly, the DNA tests show Jean to have a criminal predisposition for violence and the genetic para-counsellor advises that the state would seek preventative imprisonment in a foster home. Distraught, Struglin', who is deep in the depths of post-natal depression, slashes Jean's and her own wrists with a knife stolen from the kitchen. A chance saves Jean, but Struglin' dies, certain she would be forgiven the terrible crime she had committed against her baby to save her from an inevitable life of suffering and deprivation.

☐ **Impossible practically.** I have discussed gene differences and criminality in relation to Jeanne's life and these comments apply equally to Jean.

> Over the next 15 or so years, Jean is brought up and educated, poorly, in a state institution for the genetically underprivileged (SIGU). She suffers from dyslexia and attention-deficit disorder and remains functionally illiterate.

☐ **Possible.** There is nothing new in poor education: Jean has already set off down a cycle of deprivation that may reinforce her gene's propensities. Both dyslexia and attention-deficit disorder are 'treatable' by careful and personalized education and this would certainly give her the best chance of minimizing the symptoms, but, unfortunately, Jean's needs are being ignored by society and she is given a second-rate education. She is entering what can be a self-fulfilling loop of discrimination leading to deprivation, in turn leading to further discrimination: that she is illiterate further ensures that she is unlikely to be successful in searching for stable work. These unfavourable circumstances do not necessarily mean her genes doom her; there is simply a smaller chance that her environment will favourably modify the likely consequence of her gene differences and, in the worst case, her environment might even reinforce their detrimental effects.

> In 2035, after a military *coup d'état* and victory by white supremacists, all occupants of SIGUs are tested for non-Caucasian DNA ancestry, using a battery of 24 DNA differences. Jean has 13 DNA differences that are non-Caucasian and, because this number is greater than 50 per cent, she is forcibly expelled to the United People's Country.

☐ **Possible.** I discussed the intellectual bankruptcy of concepts of 'race' and 'race purity' in Chapter 5, but there is no reason why political or social preconceptions cannot overrule the scientific truth. An ideologically motivated and sophisticated group could, without any doubt, interpret 'race' and 'race purity' in terms of possession, or otherwise, of some appropriate DNA differences that were more commonly found in one human group than another. It would not be accurately inclusive because such tests would exclude some that should have been included, but it would be a workable approximation to 'race' identity. The idea is not so far fetched: several police forces are very interested in identifying DNA differences that could be used to identify a suspect in general terms—such as White, Black, or Asiatic—using just a simple DNA test. The science-fiction version of this is to identify what a suspect actually looks like by studying the genes that define the face's features. This latter possibility is almost certainly impossible practically, but defining a probable ethnic group to which a suspect might belong is firmly in the class of possible.

> Aged just 15 and destitute, Jean joins Embryofarms as a baby-part surrogate and is paid to give birth to a brain, a headless baby, and a hand—all made using Embryofarm's standard IVF and blocking gene therapy. The embryoids are used for spare part surgery . . .

☐ **Impossible practically.** I have discussed Embryofarms in some detail and these events are currently impossible, but this does not mean that some of the problems are not already upon us. Using a woman to produce 'spare' babies is, of course, exactly what is done when surrogate mothers are used to produce children for couples who otherwise could not have them, and this is not a rare event. Central to the acceptance of surrogacy is the idea that it is a treatment for a medical condition of childlessness, but limits of this decision are very fuzzy. If it is accepted that a person has a right to a child, why does the same person not have the right to a child of a particular type? Put another way, where is the limit of the disorder of childlessness?

A fairly simple extension to this line of thinking suggests that if surrogacy is allowable to treat the disorder of childlessness, why should it not be allowable for production of embryoids, which would be used for treatment of conditions that all of us would accept as being far more firmly in the medical domain? Perhaps we do not have Embryofarms because we do not know how to make embryoids, rather than any more profound objection to using humans as spare-part machines. This is a prospect that is appalling to me—it is treating humans as less than human—but the truth is that the moral groundwork was laid some years ago by the human fertility research community. If we have a 'right' to bear a child that extends to even doing so in another human being, then we most surely have the right to 'cure' the loss of a limb by using another human to produce it—a depressing prospect.

> . . . but the company summarily fires Jean for non-disclosure of her non-Caucasian DNA test results. She survives on social payments for 2 years, but following a government decree that all citizens must undergo compulsory DNA profiling for DNA differences likely to disadvantage society, . . .

☐ **Reality.** Firing Jean for non-disclosure of test results—albeit in an artificially inflammatory context—is already a real problem today. Genetic knowledge is often double-edged. On the one hand, your employer, knowing you had a predisposition to a lung disease, might take particularly stringent precautions against exposure to dust that might harm you; in this case, you could be helped by your employer's use of the information. Equally the employer could refuse to employ you because you were a potential medical liability; the same information is now used to discriminate against your employment prospects. If you possess information that could impact on your employer or on your insurance, can either group force you to disclose this information? Legislation could, of course, be passed banning discrimination based on gene differences, but even this option is not without its problems. What is to prevent somebody, who knows he or she has a gene difference that is likely to cause them to die at a young age, from taking out an enormous life-insurance policy for their spouse or children? Knowledge of gene differences quite clearly can work for you or against you, depending on who possesses it; establishing the limits to the use of our personal genetic information is a key development that we all must face within our societies.

> . . . Jean is declared genetically unsuitable for reproduction by an increasingly right-wing government of United People's Country and is forcibly sterilized. Her profile includes the chance of common cancers at 20 per cent, Alzheimer's at 50 per cent, major depressive illness at 10 per cent, schizophrenia at 1 per cent, and dyslexia, addictive personality trait, and attention-deficit disorder and general criminality all 90 per cent—and grounds for sterilization.

☐ **Possible.** Jean is a victim of the crudest form of eugenics that I discussed in Chapter 8. Today, gene science operates, in most cases, by a code of practice that has at its core some profound views of human existence: three major principles concern Autonomy, Justice, and Beneficence.

Recognition of each of these ethical standards would mean that Jean could never be treated as she is in her future history because Autonomy is the freedom to make a decision independently of outside pressure, Justice is most easily appreciated as respect for the values of the individual human that are enshrined in laws and declarations of rights, and Beneficence is simply doing good rather than Malfeasance (doing evil). Such principles should protect future Jeans but there is no reason to suppose that the simple existence of three principles is sufficient to ensure compliance. Even today eugenics is far from a spent force. In 1995, China drafted and ultimately passed a law that was explicitly eugenic in intent and justified this on economic and social grounds—the incidence of genetic disorders amongst, in particular, minority and poor regions of China was becoming an intolerable burden upon the state.

> Jean lives a very poor existence, surviving on her wits, some drug-dealing, and stealing, and when times are hard she sells herself to men at the local rail station. Her addiction to drugs and alcohol increases and she is frequently imprisoned for her activities.

☐ **Reality.** Jean is now a member of an 'underclass', a group that has both failed within and been failed by society. Ill-educated, often diseased or with mental problems, always poor, these individuals are caught in a trap of social deprivation that is hard to escape; similar groups are to be found within all highly developed and wealthy societies. Jean's social circumstances are deeply unpleasant and oppressive, and, if reality is unbearable, alcohol or drugs are an escape—her environment might well reinforce any of the traits contained within her gene differences. Even in these dire circumstances, however, there is no certainty: the chance that she will become an alcoholic may be increased, but it is still not 100 per cent. She is not necessarily doomed for, even in these extremes, genes still do not dictate.

The occupants of the underclass are a target for eugenic thinkers simply because they are at the margins of society, are often financially supported by the state without themselves making financial contribution, and often are disproportionately involved in crime. The point that even in these extremes of social deprivation, outcome is as much affected by social conditions as by gene differences is lost on people who wish to have a simplistic and inaccurate view of the determining nature of genetic variation.

She is regularly treated for a whole range of sexually transmitted diseases using twentieth-century antibiotics, which are predominantly ineffective because of multiple resistances in the disease organisms, and finally becomes infected with HIV. The state refuses her any further treatment because her DNA profile suggests that the incubation period before AIDS becomes life-threatening is greater than 4 years and therefore deemed to be outside of the 3-year minimum-life expectancy period policy (3MEPP) that triggers treatment under the medical rationing programme for the genetically disadvantaged started in the early twenty-first century.

☐ **Reality and possible.** Multiple drug resistance is an increasingly dangerous problem: at present, we are seeing just the iceberg tip of the cases. The origin of these unpleasant bacteria is rather unusual and a classic result of evolution at work. There are literally millions of different species of bacteria— we have no idea how many to be frank—and each species has become highly refined in the environments it can occupy. These can be extraordinarily harsh: the frozen snows of the Antarctic or the superheated water columns near thermal vents deep under the sea. To enable bacteria to survive, they have developed a huge range of proteins that can deal with the countless millions of different chemicals they meet in their chosen worlds. So when we humans make an antibiotic that is based on a chemical found in nature, we can be pretty sure that somewhere in the world is a bacterium that makes a protein that destroys the antibiotic: this was exactly the case for penicillin, a natural product of a mould that is destroyed by the protein called penicillinase.

Why should bacteria have penicillinase? Precisely so they can live next to a mould that produces penicillin, and so occupy an environment where other bacteria cannot venture. Such is the ferocious competition in the bacteria's world that the genes that make these sorts of antibiotic-destroying proteins are of huge value to bacteria, and so they have evolved ways of passing the necessary genes between them. They achieve this feat by collecting up many of the genes that confer antibiotic resistance onto a single DNA fragment and passing this between different members of the same species to make sure they can all live in the same otherwise hostile environment. If a bacterium is given one of these DNA fragments, it will be able to make several new antibiotic-destroying proteins. The result is that it will simultaneously become resistant to several antibiotics; this is the classic case of multiple drug resistance. The ability to pass the resistance genes from one bacterium to another of the same species and even sometimes to a different species means that multiple drug resistances spread rapidly and unpredictably.

To fight the problem of drug resistance, we urgently need new antibiotics, based on wholly novel chemical compounds; this is where the genome projects that I discussed earlier will become a vital source of new antibiotic targets. The complete list of proteins that bacteria can make provides a wonderful start to developing new antibiotics. This is not only possible gene science, it is near reality: the next-generation antibiotics, developed on this type of knowledge, are being tested as I write.

Alarmingly, however, targeting of treatment is becoming a fact of public health services, and the spiralling costs of modern drugs, combined with ever-tightening fiscal control by governments, mean that this will probably become more, not less, common. There is no doubt that gene differences are increasingly useful in classifying medical conditions and classification is, of course, centrally important to medical judgement on 'survival', 'severity', and all of the other clinical consequences of disorders. This means that gene differences are indeed being used to define a likelihood of outcome and therefore inform treatment and advice; they are a reasonable, indeed desirable, clinical aid. But it is just a short step to using the same information as an aid to avoid treatment, for simple reasons of cost and not to aid clinical judgement. Bearing in mind that gene differences do not bring certainty, only possibility, this would be a very dangerous development indeed.

> Jean is now aged 26 but looks like a 56-year-old, her good looks having gone. In 2048, she is diagnosed as having a slow-growing brain tumour, which causes unpredictable blackouts. This is not treated under the 3MEPP. Sleeping in a derelict building, she catches TB and is found by the police, collapsed and half dead on the street, and is taken to the poor hospital where she is isolated from all human contact because she is harbouring a multi-drug-resistant TB strain. Why she does not die, no one can tell: but she doesn't.

☐ **Reality.** Multiple drug-resistant TB bacteria are a real and dangerous problem right now, particularly in the old Soviet Block, where the breakdown of public health services has had a very marked effect on public health. TB affects about 8 million people annually and is commonly treated with a combination of up to four drugs—isoniazid, rifampin, ethambutol, and streptomycin. In about 1.4 per cent of cases the disease-causing bacteria are resistant to more than one of the drugs. Compounding this problem, the TB bacteria spread rapidly and easily from person to person and so treatment requires physical isolation of the patient to prevent infection of nursing staff and other patients. Given that the drug treatment may often take 6 months or more, this is a devastating prospect for most individuals.

> She finally emerges from hospital in late August just in time to be caught in the cloud of infectious particles released from a missile containing a viral warfare agent that has been fired from the adjacent United People's Country. The agent had been genetically modified to target only non-Caucasians and 5 days later, in great pain and blind and deaf from the nerve destruction caused by the deliberately designed tissue specificity of the warfare agent, Jean dies—on the last day of August 2048.

☐ **Impossible practically.** I have discussed the difficulties of producing an ethnic weapon in Chapter 5; an ethnic weapon would need a protein 'target' that is found only in one group and not in another and that was a component of the infectious route of the warfare agent. At present, there are relatively few known potential targets that could be used as virus entry points into human cells and none of these, to my knowledge, has changes specific to any ethnic group. What this means is that we have no evidence whatsoever that a weapon could be tailor-made, as was the one that killed Jean—this is in the realms of the impossible practically. It is quite possible to think of ways a virus might be built that could destroy nerve tissue in quite a specific way—Jean's death is in this respect possible—because gene engineering can be used to make viruses that have new and malign properties, much as viruses can be given new and benign properties for gene therapy. I would be surprised if such modified viruses did not already exist in some laboratories around the world, but they would be able to infect or kill all or most humans rather than a specific ethnically defined subset—not a particularly comforting thought, I grant you.

> Jean lived and died in the future, where disease and suffering were the results of nature and malign human influence.

☐ **The reality.** The truth is that Jean Battler suffered at the hands of her genes in a way that Jeanne Dream did not for two quite separate reasons. Gene differences can inflict terrible suffering and pain on some individuals and this, regretfully, is a simple fact of existence; some of Jean's suffering was a consequence of bad luck. She could have been even unluckier and born with a crippling disorder such as muscular dystrophy, a lingering condition such as cystic fibrosis, or any one of the 10000 or so disorders that are believed to be associated with gene differences. So some of Jean's fate is simply to have been given the wrong gene differences at the moment she was conceived.

What was avoidable in Jean's unhappy life was the suffering that was caused by the abuse of the knowledge of her gene differences. If you compare the courses of Jeanne and Jean's lives, then it becomes obvious that knowledge of gene differences can be used to inform treatment or the same knowledge can be used to stigmatize, discriminate, and to condemn. The clear implication is that knowledge of gene differences is not, in itself, dangerous to an individual; it is the uses to which this knowledge is put that will ultimately define it as benign or malign. If you accept this view, we should not be afraid of genetics because it can be a servant to us, but perhaps we should tightly control the access to such knowledge. It is the fear of genetics that is the final topic of this book.

So what are you afraid of now?

In this book, we have covered most of genetics, and in the future histories of Jeanne and Jean we have explored the future. We have as much information as we need to make reasoned judgements on the reality of genetics, on its power, its limitations, and on its potential. So I want to pose one final question: does gene science frighten you? Most people will generally admit to an element of fear in their attitude. From personal observation, there seem to be three main areas that cause this: the fear of knowing the future, the fear of the loss of free will and individuality, and, finally, the fear of discrimination. Let's take each fear in turn and see what our knowledge of genetics makes of the reality behind the concerns.

Knowing the future

The fear of the future is intimately tied up with our fear of death and disease and is based around the view that gene differences can be used to predict future health and likely cause of death with certainty. If gene differences dictate what will happen to you, a knowledge of gene differences is identical to a knowledge of the future and this is the source of our fears: we do not wish to know the cause or the time of our death.

We have seen that certainty is not an effect of gene differences; gene differences simply suggest what is likely to happen and not what will happen to a person, and the distinction between 'likely' and 'will' is extremely important. No amount of DNA analysis will ever be able to produce certainty of outcome because that is the nature of our genes and their relationship with our environment. This does not mean that the fear is groundless; genes can make predictions based on likelihoods. So even though we could never be certain, we could at least have some indications as to our future fate and this information can be very damaging. In some senses, uncertainty is almost more difficult to deal with than certainty; at least certainty provides a solid fact that can be faced and embraced whereas a likelihood always will provide an avenue down which we can escape—perhaps it will not happen, perhaps the 80 per cent chance will mean that we are among the 20 per cent of survivors.

Each of us deals with the fear of our Genetic Destinies differently. Some may welcome the knowledge and others may hide from it, and no one has the right to argue that one position is more acceptable than the other. This view underpins much thinking about genetics and its application to human beings. One key concept has evolved to try and ensure these fears are met; the concept is elegantly simple—it is the right of ignorance.

Every person has an inalienable right to remain ignorant of their gene differences because this is the only way that individual dignity and respect can be maintained; it is a cornerstone of thinking about the social impact of genetics. The fear of a certain future is not based on the reality of genes, but the fear of possible futures most certainly is: it is a new kind of knowledge about which we have to be educated and with which we have to become familiar. If we can do this, then some of the fear will be replaced with hope, because with the knowledge also comes the possibility of avoidance.

The end of free will and individuality

'It's in my genes'—the phrase that is the tacit recognition that our lives are beyond our control. If gene differences define character, intelligence, and health then we cannot escape our Genetic Destinies and we no longer have free will: we are no longer individuals, we are automata, controlled by our genes. Very clearly, this view is not based on genetic reality at all. Every one of our discussions has pointed to the interconnection of gene differences and environment in determining our futures, and our actions must, therefore, determine our futures as much as our gene differences.

Nevertheless, the fear persists because even some genetic influence can be perceived as degrading our ability to act as a free agent in the world. Let's look at this a bit further. Genes are deterministic in the sense that our genes make us human, but gene differences make us uniquely unlike any other person in this world and this uniqueness is reinforced by the equally distinctive environmental influence on each of us. Does even this amount of influence mean we have limited free will? This becomes a question of degree: if you have infinite choice, you may consider yourself to be free, but what if you have a billion, billion, billion, billion choices? Or a million? Or ten? Or one? When does the limitation of free will become reality? To me, the whole argument over free will is bogus and is not based on a biological reality. We are limited in our free will because we are human: I cannot act or think like a tree, a dolphin, or a worm, because I am human, which could be taken to mean I lack free will. Most people would accept that this is stretching the meaning of free will past breaking point.

Where, then, is the actual limit of our free will? It is in the combined weight of our gene differences and our life experiences; this must be a bounded freedom but these are bounds of such magnitude that were we to live a billion lives each for a billion years we could never experience all of the potential states that our gene differences and environments would allow. This is not infinite freedom but it means that we need have no fear of our genes' limitations on our lives. We must recognize that in our lifetimes we will be able to explore only an infinitesimally small part of all that there is to be explored. Perhaps this is not 'free' will, but it is so close that we will never see the difference.

Discrimination and the inescapable life sentence

The gene differences in each of us can be used to help treat and avoid disorders and diseases, and most people would see this, within the limits we have discussed, as of tremendous potential medical benefit. As the stories of Jeanne and Jean have graphically illustrated, however, the information contained in our genes can equally be used to discriminate against us. Discrimination takes many forms and we face three main areas of potential difficulty: racial, behavioural, and medical discrimination. In all three cases discrimination is, once again, based on the untenable view that gene differences can be used to make absolute statements about the person concerned and this, we have seen, is only very rarely the case.

Racial discrimination

I have discussed the concept of 'race' in some detail and, in a sense, this is perhaps the easiest problem to deal with from a scientific, if not a political, viewpoint. The case is quite clear: genetics does not and cannot provide the objective underpinnings for a racial classification of human beings. We can be sure that this minor inconvenience of fact will not stop some people from attempting a spurious use of genetics to support racial prejudice. Perhaps the most important thing we can do to resist this abuse is to identify the motives behind attempts to classify human beings, and then approach the problem as a political, rather than a scientific, problem. Ultimately the solutions to racial antagonisms lie well outside genetics and it is the task of professional geneticists to make sure that their science never again becomes sucked into the support of explicitly political agendas.

Behavioural discrimination

Discrimination based on behaviours are far more problematic to genetics because such discrimination is already common within our societies. I briefly discussed (in Chapter 6) the difficulties presented by gene differences such as that occurring in the MAO gene and the propensity for violent behaviour that seems to be associated with it. How should we respond to the discovery that a person has a gene difference such as this? Should we assume that he or she would inevitably become violent and so preventatively lock them away before they can do any damage, as we do with some mentally ill individuals? I think this is a dangerous approach because, once again, we are confusing certainty of genetic outcome with likelihood. Given gene differences do not determine, only influence, knowledge of possession of a gene difference cannot then become grounds for supposing that a particular behavioural outcome is certain to happen: natural justice dictates that we have to continue to assume that an individual is innocent until events teach us otherwise.

The confusion behind our thinking in this area is perhaps best illustrated by an example taken from outside the field of genetics. We know that young males are involved two or three times more commonly in accidents than older males or any female; indeed, the leading cause of death in the young is road accidents. So would it not be a reasonable proposition to lock up all young males to avoid the damage and death? This is clearly absurd because we have missed the point that even though road accidents are common in these males, most young men are not involved in road accidents at all. Once again, we have confused certainty and likelihood.

How can we proceed through this difficulty? One possibility is simply not to allow gene differences to be used in a court of law or to determine society's actions. There is, of course, a cost to this—some violent individuals will surely commit violence—but it is a cost with which our societies are already familiar, albeit in a new context. Natural justice surely dictates no other course of action? The inadmissibility of genetic evidence would equally apply to attempts to excuse violence by apportioning blame to gene differences; the excuse that 'it's in my genes' is base on complete genetic determinism and so does not correspond to scientific reality. By this argument, a person is responsible for their actions quite independently of their genes and this seems a logical and accurate position.

Medical discrimination

Medical discrimination has become a particularly contentious issue because of the health and life-insurance and assurance industries. Life-assurance companies have been making likelihood of death calculations for years and have developed simple ways of allowing for an increased chance of death; smokers, for example, often pay an increased amount for the same cover as a non-smoker or an individual may even be refused cover altogether if he or she insists on leading a dangerous life full of Himalayan mountain-climbing and scuba cave-diving. Why, then, should a prospective insurer not insist on a full genetic test to seek out those likely to die or become ill? After the discovery of a gene difference predisposing to a life-threatening condition, an individual must be perceived as being a poorer risk for life assurance—or housing loans or, indeed, any long-term investment by an employer. This hides a very difficult problem: what of those that do not pass such a test? Are we simply going to allow a proportion of our population to be excluded from the benefits of insurance or a housing loan or even employment?

Perhaps we can make the callous argument that this is a problem only for a tiny minority and we don't, therefore, have to worry. This unpleasant suggestion is deeply flawed but thinking about it offers perhaps the way out of these difficulties. One fact of existence is incontrovertible: we are all going to die of something. Eighty per cent of us will die of one of the top 10 killers of human beings, eight of which have a significant contribution of gene differences; and this, in turn, suggests that probably all of us will contain some gene differences that may contribute to our death. As genetic knowledge of the common gene differences becomes more complete, more and more genetic tests will result in more and more of us becoming excluded from insurance, loans, and employment. This is absurd; in time all of us will be excluded, leaving no industry and no one to employ! What is happening now is that we are presently in a state of partial knowledge: there are relatively few gene differences known that contribute to human mortality and so possession of one of these differences is indeed exceptional information, but the underlying, if as yet undiscovered, truth is that we are all at genetic risk.

Can we dig ourselves out of this absurd position? One solution is simply to make it law that no employer or insurance company can request a genetic test and the results of such tests cannot be disclosed to anybody except the doctor and the individual tested. This is a very reasonable idea and the right to genetic privacy is becoming an attractive concept in thinking about this area. The recent decision in the UK to allow the insurance industry to use the results of some genetic testing is an unfortunate development, and it is unclear how the authorities will subsequently modify this position to avoid the absurdity of wholesale genetic discrimination that will inevitably result. The even more recent announcement (November 2001) of a 5-year suspension of such testing is laudable but the potential for confusion remains unrecognized.

The observation that both sides can use genetic knowledge—the insurance industry and the individual—has caused much fuss recently. The concern is simple; what is to stop a person who has had a genetic test and discovered that he or she has a gene difference that is likely to kill them at a young age from taking out an enormous life-assurance policy, knowing that their children will benefit at an early date? In this case, possession of the genetic information by the customer gives them an unfair advantage over the insurance industry. It's not clear how real a problem this might be because we must still keep in mind that gene differences are not deterministic and so even though an individual might well have an increased likelihood of dying young, this is not a certainty; they might well live to a ripe old age and have to continue to pay premiums like the rest of us. The solution to this problem is most simply within the hands of the industry: they need to be informed enough to know how common are the gene differences they are concerned about and they can then adjust all of their customers' premiums to take account of the proportion of frauds that are likely to occur—the calculation is one that has been explicitly or implicitly carried out many times before.

We cannot pretend that knowledge of gene differences is easy knowledge: there are certainly consequences that are going to need a careful social response and I have just touched on only some of the key features of the current debate. We can, I think, be optimistic that widespread discrimination is not going to be a feature of our new knowledge, but we must also be realistic enough to believe that there will be cases of abuse and the law will have to be changed to deal with these. We are perhaps in a similar situation to laws against murder; the existence of such laws cannot guarantee that people will not murder each other, they simply define punishment. Perhaps this is what we need to control the use of our most personal genetic information—a law to punish transgressors rather than a law to prevent transgression.

A Genetic Destiny

We all have a personal destiny—the path of our life—but we have no way of predicting what it will be; there are moments when we control our destiny, when we choose partners, jobs, careers, but most of our lives are led in an exciting mix of plans and reactions to events. Our genes and the differences they contain have moulded us from the moment of conception, but, as our lives unfold, we face a myriad Genetic Destinies as our genes and our lives combine in an intricate dance of mutual influence. So this is our true Genetic Destiny: to live our lives as human beings, unique, unpredictable, and irreplaceable in all of history and all of future time: it is a wonderful prospect.

The yrassolg: a reverse glossary

This is the key to making a simple subject more complex! The reverse glossary provides the technical names for some of the features I have discussed. I have made no attempt to make this comprehensive—many terms will remain simplified. Instead, I have selected those terms that are commonly found in general discussions of genetic results in press announcements and the like.

accident This is simply used by me to describe an event that happens by chance, rather than an event that happens for a specific reason. So it is accidental that two people have the same DNA difference when they are unrelated; *see* chance, random.

AGCT The collective name for the four 'nucleotides' in DNA: each is named after the chemical 'base' it contains, so A = adenosine (after the base adenine), G = guanosine (guanine), C = cytosine (cytidine), and T = thymidine (thymine).

B and C mice The real names are DBA and C57/Black6; these are simply designations of mouse strains.

back and front The correct term for these in developmental biology is 'dorsal' and 'ventral'.

cell division This is called 'mitosis' when normal cells divide, but when it is the special division associated with egg or sperm production, where the new cell will be given only one set of genes, it is called 'meiosis'.

cell-growth enhancers There are a huge number of these so-called 'growth factors'. The reason they are called 'factors' is because the growth-enhancing activity was experimentally identified without knowing what was responsible for it; usually they are proteins but sometimes more simple molecules can be growth factors.

chance, random The correct term is 'probability'; being hit by a meteor is a low-probability event, but taking a breath within the next 10 seconds is a high-probability event. When the value can be mathematically quantified, we give a figure, so the chance of tossing a coin to get 'heads' is one-half or a probability of 0.5; and tossing it twice to get two heads is 0.5×0.5, or 0.25.

code words The three sequence of three rungs that code for an amino acid is correctly a 'codon'.

copying DNA In technical language, DNA is a polymer—it is made up of many identical or near-identical chemical components (such as the sugar and phosphorus atoms in the 'rails'). So making more DNA is an example of 'polymerization' and the proteins that do this are correctly called 'DNA polymerase'. Proteins that cause a chemical reaction are given the suffix '–ase', hence polymerase. DNA polymerase in humans is a complicated multi-protein structure.

copying DNA into RNA The process of making a half copy of a gene into messenger RNA is called 'transcription'. This is because the DNA code is being written (transcribed) into RNA that will then be translated into the correct order of amino acids in the final protein. The complex of some 40 proteins that transcribes DNA into messenger RNA is called 'RNA polymerase'; *see* copying DNA.

differences (changes) in DNAs The correct term is 'mutation', whether the change has a functional consequence or not; mutation simply signifies that a different chemical rung is found in the same place in two or more people. A distinction is sometimes drawn between 'variation' and mutation, with variations occurring in many people and mutations in only a few, but the distinctions are not well drawn. Increasingly variation is replacing mutation in discussions over human differences because of the sensitivities of the people who possess the change.

environment The term is used correctly in the book to mean everything that is not connected with genes and could influence our lives; this includes what we learn and see and sense as well as more physical events such as illness, poisons, radiation, and so on. The word should not be confused with the narrower use in terms such as environmentalist.

fragments of DNA (23) These are usually called 'chromosomes'. In humans, each of the 23 chromosomes is neatly and separately wrapped up in a protein coat and the combination of protein plus DNA is readily stained by dyes and made visible under the microscope (hence the name, which means 'coloured body'). The DNA fragments are numbered from 1 to 22 and, in addition, there is a special pair of 'sex' chromosomes, X and Y. The largest chromosome (1) has about 300 million rungs and the smallest (21— it should have been number 22 but the original studies were wrong) has about 30 million rungs. Strictly speaking the DNA in a chromosome is not synonymous with a chromosome, which consists of protein as well as DNA, but general usage means that the same word frequently applies.

gene difference *See* differences in DNAs. The term is used to apply to any DNA change within a gene or its controlling DNAs.

gene regulator The correct name is 'transcription factor'. Messenger RNA is copied from DNA in a process called 'transcription', hence the first part of the name. Why factor? Because in the earliest stages of the research to isolate the first transcription factor, no one knew what was contained in the samples as they were purified, and so they were called a 'factor' rather than guessing that it was a single protein. Many, but not all, transcription factors are made up of several different proteins or of more than one copy of the same protein.

half ladder Because it has only one 'strand', it is referred to as 'single-stranded'. It can be made of either DNA or RNA; if it is the messenger, it will be RNA.

head and rear The correct developmental term is 'anterior' and 'posterior' if you mean towards the head or rear end, but if you really mean 'head' and 'rear' you could say 'rostral' and 'caudal'. It is very confusing for all except developmental biologists, who seem to thrive on obscure and ambiguous terms such as these.

identical twin The correct name is 'monozygotic' twin. The 'zygote' is the newly fertilized egg and so this implies that the twins derive from the same zygote. The splitting of the zygote can happen after the fertilized egg has started to divide, however, so this is strictly speaking not a completely accurate term.

ladder of DNA The correct term is the 'double helix' and, because it normally contains two rails, it is normally 'double-stranded'.

non-identical twin The correct term is 'dizygotic' twin because each is formed from a different zygote, the technical term for a fertilized egg; *see* identical twin.

one gene, a few genes, or lots of genes The correct terms for these generally refer to disorders caused by changes to one gene (a 'monogenic' disorder or condition), to a few genes (an 'oligogenic' condition), or to lots of genes (a 'polygenic' condition). The original meaning of this last term has itself become changed over the years. Originally a 'polygenic' condition was caused by so many gene changes that theoretically none of them could be detected by linkage analysis. There may well be polygenic conditions in the old sense of the definition—some behaviours may fall into this class—but its use nowadays tends to be in place of oligogenic. Scientists are not always strong on the history of their science!

partner (protein) The correct name is 'ligand'. A ligand can be any chemical and does not have to be a protein; for example, when you taste food, chemicals in the food you are eating are the ligands for a whole group of taste signalling proteins. In most cases in this book, the ligands are indeed partner proteins.

protein functions Many proteins, functionally, are 'enzymes'. Enzymes induce chemical reactions that might either never happen or happen too slowly to be useful in a cell. Without enzymes all life would be impossible because all of the myriad reactions that go to make a cell would never occur fast enough to enable normal function.

rails The technical term is 'strands', so DNA is generally 'double-stranded'.

rungs These are 'nucleotide base pairs' and each base pair gets it name from the AGCT chemicals—the 'nucleotide'—that are their major component; *see* AGCT.

signalling protein The correct term is 'receptor' protein. A receptor protein is generally located on the surface of the cell with an internal region that can 'signal' to the interior once its partner protein has stuck to it on the outside. The name is accurate—it indeed 'receives' the partner—but it does not convey the importance of the subsequent signalling actions, hence my use of the term 'signalling protein'.

stuck, stickiness The correct terms are 'bind' (or 'binding') and 'affinity'. A partner protein has a strong affinity for its signalling protein and so binds to it. Many of the complicated interactions in living cells are caused by proteins binding to each other; this can be a very specific process because, like a lock and key, the 3-D shape of each protein allows a precise fit. The affinity of two proteins for each other is frequently both highly specific and very strong.

switch DNA The correct term is complicated because there are actually different sorts of 'switch DNA'. One type is called a 'promoter' and corresponds to a region of DNA, generally near a gene, that controls the specific rung location where the production of messenger RNA will start. In contrast, regions, often distant to the gene (sometimes many hundreds of thousands of rungs away), contribute to the control of the time and the place when and where a gene makes messenger RNA; this type of switch DNA is an 'enhancer'. To add a complication, promoters may also have enhancer-like functions. In general, the control of genes in humans and other complicated, multi-celled creatures is much more complicated than in the single-celled bacteria, where enhancers do not feature.

variable regions of DNA There are several correct terms for DNA whose rung sequences differ slightly in length; these includes VNTR for 'variable number of tandem repeats' (a tandem repeat is AGCT ACGT ACGT, where ACGT is the tandem repeat). The repeating DNA sequence can consist of two, three, four, or more rungs and so these are referred to as 'dinucleotide' (two-rung), 'trinucleotide' (three-rung), or 'tetra-nucleotoide' (four-rung) repeats. 'Minisatellite DNA' and 'microsatellite DNA' have longer repeats. Some special DNA sequences are simply 'hypervariable' and are very different for no obvious reason.

Index